INSTRUCTOR'S RESOURCE MANUAL

KATHERINE M. WHATLEY ✦ JUDITH A. BECK
The University of North Carolina at Asheville

and

JUST IN TIME TEACHING NOTES

GREGOR NOVAK ✦ ANDREW GAVRIN
Indiana University - Purdue University, Indianapolis

PHYSICS

Second Edition

JAMES S. WALKER
Western Washington University

Pearson Education, Inc.
Upper Saddle River, NJ 07458

Associate Editor: Christian Botting
Senior Editor: Erik Fahlgren
Editor-in-Chief, Physical Sciences: John Challice
Vice President of Production & Manufacturing: David W. Riccardi
Executive Managing Editor: Kathleen Schiaparelli
Assistant Managing Editor: Becca Richter
Production Editor: Dana Dunn
Supplement Cover Management/Design: Paul Gourhan
Manufacturing Buyer: Ilene Kahn
Supplement Cover Designer: Joanne Alexandris

Pearson Prentice Hall
Pearson Education, Inc.
Upper Saddle River, NJ 07458

Printed in the United States of America

10 9 8 7 6 5

ISBN 0-13-140659-0

Pearson Education Ltd., *London*
Pearson Education Australia Pty. Ltd., *Sydney*
Pearson Education Singapore, Pte. Ltd.
Pearson Education North Asia Ltd., *Hong Kong*
Pearson Education Canada, Inc., *Toronto*
Pearson Educación de Mexico, S.A. de C.V.
Pearson Education—Japan, *Tokyo*
Pearson Education Malaysia, Pte. Ltd.
Pearson Education, *Upper Saddle River, New Jersey*

Instructor's Resource Manual

Katherine M. Whatley
Judith A. Beck
The University of North Carolina at Asheville

---

to accompany

Physics

Second Edition

James S. Walker
Western Washington University

Instructor's Resource Manual
and
Just in Time Teaching Notes

to accompany *Physics, second edition* by James S. Walker

Table of Contents:

Instructor's Resource Manual

to accompany *Physics, second edition,* by James S. Walker

Table of Contents

Introduction to the Instructor's Resource Manual

This Instructor's Resource Manual for *Physics, second edition,* by J. S. Walker is aimed primarily at instructors who are relatively new to teaching or new to this course. We hope it will be helpful for more experienced instructors as well. Both of us have had many years of experience with algebra- and trigonometry-based physics courses. We present here ideas and suggestions, not a blueprint for the "right way" to teach. We view teaching a course as a cooperative enterprise that actively involves the students and the instructor. We hope you find this manual helpful as you develop your own unique approach to teaching and learning.

Physics for the Non-Major

Most of the students who enroll in an algebra- or trigonometry-based physics course are not planning to be physics majors. Some will be majoring in other sciences (biology, environmental studies, etc.) and some will be non-science majors. These students often bring fear of physics to the course. It is important to address these fears early and to reassure the students that physics is something they can learn. It is also important to discuss why physics is important to them. Physics provides ways of modeling and understanding the behavior of the universe; if we can model behavior, then we can predict it and determine how to influence it. How well we can predict behavior depends on the quality of the model and on how well information can be known. Most of these students will have a Newtonian worldview. Discussions of how well information can be known lead into discussions of the quantum mechanical worldview, which many students find confusing but interesting.

Discussing physics as a way of modeling the universe also gives justification for some of the simplifying assumptions we as instructors take for granted. Why do we neglect air resistance when talking about objects falling near the surface of the Earth? Because it makes the model simpler and easier to understand. We know that in later courses air resistance will be added into the equations to build a more complex model. Did the first automobiles have all the extras that modern cars have? No – at first it was enough to build a car that ran and didn't explode. Later, after the simple model worked in a predictable way, engineers could add additional details (windscreens, windows, streamlining, heaters, fuel injectors!), and the car evolved into a very complex machine.

Examples and demonstrations are critical for these students. The more concrete, everyday, visual examples you can provide the better. It is also useful to tailor the examples to the class. If most of the students in your class are interested in the health professions, biological examples that involve the human body are appealing. If the class is full of literature or history majors, bring in examples of the writings of famous physicists or examples of when physics was important in history.

Course Organization and Structure

Physics will often not be the most important course non-majors will take in a given semester, at least from their point of view. It is advisable to provide sufficient structure so that the students know where they are in the course, know what will happen and when, and are encouraged to stay up to date. It is very helpful for students to have a complete syllabus, including homework assignments and test dates, at the beginning of the semester. This helps them plan out their semester and enables them to inform the instructor in advance if there are scheduling conflicts.

Each class has its own rhythm. As the instructor, you can set the rhythm by providing a variety of formats throughout the class to keep students alert and involved. For example, the class could begin with a short quiz over the previous class material, then proceed with a question-and-answer session on the homework, a demonstration illustrating the concepts to be covered that day, a discussion of the concepts, and a series of examples. Conclude with an active learning experience (doing a short experiment or working an example in teams). Students generally like structure and being able to anticipate what will happen next. Do remember to be flexible and look for "teachable moments," such as questions from the class that lead into discussions pertaining to the concepts of the day.

Problem solving can be very difficult for beginning physics students. Many are not familiar with the approaches and techniques of physics, and others feel their math skills are a little rusty. Consider designating one of your office hours per week as a problem session. Schedule it one or two days before homework is due each week, and hold it in a classroom or seminar room instead of your office so that you can circulate among students as they work. This approach allows you to help many students at once, including some who might be reluctant to come individually to your office hours. An added benefit is that students will help each other, and as teachers, we recognize that one of the best ways to learn something is to try and teach it to someone else!

There are a variety of ways to encourage students to keep up with the class. Collecting **homework** every week is quite effective; it gives the instructor immediate feedback about which concepts were not clear to the students, and it provides the students with incentive to practice problem solving. Homework is most effective when at least a portion of it is graded and returned at the next class, or when solutions are discussed as soon as the homework is collected. It is generally not effective to assign massive amounts of homework, collect it, and never give the students feedback on it. The students tend to get discouraged and stop doing homework at all. Posting solutions on the Web, on reserve in the campus library, or on a bulletin board will allow you to provide complete explanations without spending a lot of class time on solutions.

Quizzes provide another way to encourage students to stay engaged and caught up. A relatively simple and short quiz given at the beginning of class once or twice a week can ask the students questions about the last class (which encourages them to look over their notes) or questions about the material to be covered that day (which encourages them to read the material before coming to class). Alternatively, use technology and give a quiz via computer that must be completed before class. The in-class quiz can also be used as an attendance check and as an incentive to get to class on time; the Web quiz allows you more class time for discussion and examples.

It is a good idea to give a longer quiz that requires problem solving every few weeks and at least once before the first test. No matter how much you tell students to practice, those who have not taken physics before may not understand the critical difference between understanding a problem solution when you write it on the board and being able to do it themselves. Bombing the first quiz is a great learning experience and won't affect the overall grade too much in the long run. Bombing the first test is a much more costly learning experience.

Tests are an important component of a physics class. We recommend giving three per semester or two per quarter (instead of a single midterm), and a comprehensive final exam. Although it's hard to give up class time to tests, we find that students benefit from frequent evaluation, especially the first semester or quarter when they are learning *how* to study physics in addition to learning the content. Include some conceptual questions on the tests as well as problems, particularly if you have been emphasizing conceptual understanding in class. These can include ranking exercises, qualitative sketches of diagrams or graphs, short explanations, and multiple-choice questions. A test format with about 25 percent conceptual questions and 75 percent problem solving is one that works well. Beginning physics students often worry about memorizing all the formulas they need to solve problems. Consider allowing them to bring an index card of formulas for use on the test. This will enable them to concentrate more on *how* to solve problems and less on memorization. Be very specific about what is allowed on the card, and point out to them that after all the practice they are doing, they should be quite familiar with the formulas and will probably know them anyway.

Consider using a variety of **evaluation methods**. It is best for student learning to have a variety of ways to earn points and not to have a majority of the course grade dependent on the final exam. How the instructor structures the grading for a course depends on the instructor, the number of students, the presence of a lab component, and the institutional culture. It is critical to be consistent, fair, and prompt with feedback. Students need to know at the *beginning* of the course how they will be graded. Several possible grading schemes follow.

Scheme 1: Semester course with lab

Labs 5%; Homework 10%; Quizzes 5%; Tests (3 @ 20%) 60%; Final Exam 20%

Grading scale: A: 92–100, B: 82–90, C: 72–80, D: 62–70, F: 0–60

In this scheme, a gap is left between the letter-grade ranges for the discretion of the instructor. Often, borderline cases are decided primarily by the student's performance on the final exam.

Scheme 2: Semester course with lab

Labs 6%; Homework and Quizzes 15%; Tests (3 @ 18%) 54%; Final Exam 25%

Grading scale: A: 90–100, B: 80–89, C: 70–79, D: 60–69, F: 0–59

Scheme 3: Quarter course with lab

Labs 5%; Homework 10%; Quizzes 5%; Tests (2 @ 25%) 50%; Final Exam 30%

Grading scale: A: 92–100, B: 82–90, C: 72–80, D: 62–70, F: 0–60

In this scheme, a gap is left between the letter-grade ranges for the discretion of the instructor. Often, borderline cases are decided primarily by the student's performance on the final exam.

Scheme 4: Semester or quarter course without lab

Homework 15%; Quizzes 10%; Tests 50% (3 per semester or 2 per quarter); Final Exam 25%

Grading scale: A: 90–100, B: 80–89, C: 70–79, D: 60–69, F: 0–59

Sample schedules

Suggested times to spend on the material in each chapter for both a two-semester and a three-quarter course are given in the following tables. We present schedules for two formats: three 50-minute class meetings per week and two 75-minute class meetings per week. The schedules assume 15 weeks per semester or 10 weeks per quarter, for a total of 30 weeks for the entire course in either case. The order of topics is different from the textbook for the three-quarter course to avoid splitting the electricity and magnetism section between quarters. Note that the schedule includes time for three tests per semester (or two tests per quarter) instead of a single midterm exam. The schedules below assume a separate final exam period at the end of the semester or quarter, during which you will give a comprehensive final.

SCHEDULE FOR A TWO-SEMESTER COURSE: FIRST SEMESTER Mechanics, Thermal Physics		
Topic	**# of 75-min. lectures (2/week)**	**# of 50-min. lectures (3/week)**
Ch. 1	1	1
Ch. 2	1.5	2
Ch. 3	1	2
Ch. 4	1.5	2
Ch. 5	2	3
Ch. 6	2	3
Test #1	1	1
Ch. 7	1	2
Ch. 8	2	3
Ch. 9	2	3
Ch. 10	1.5	2
Ch. 11	1.5	2
Test #2	1	1
Ch. 12	1	2
Ch. 13	1.5	3
Ch. 14	2	3
Ch. 15	1.5	2
Test #3	1	1
Ch.16	1	2
Ch. 17	1.5	2
Ch. 18	1.5	3
TOTAL	**30**	**45**

SCHEDULE FOR A TWO-SEMESTER COURSE: SECOND SEMESTER Electromagnetism, Light and Optics, Modern Physics		
Topic	**# of 75-min. lectures (2/week)**	**# of 50-min. lectures (3/week)**
Ch. 19	2	3
Ch. 20	2	3
Ch. 21	3	5
Test #1	1	1
Ch. 22	2	3
Ch. 23	3	4
Ch. 24	2	3
Test #2	1	1
Ch. 25	1.5	2
Ch. 26	2.5	4
Ch. 27	1	2
Ch. 28	2	3
Test #3	1	1
Ch. 29	1.5	2
Ch. 30	2	3
Ch. 31	1.5	3
Ch. 32	1	2
TOTAL	**30**	**45**

SCHEDULE FOR A THREE-QUARTER COURSE: FIRST QUARTER Mechanics		
Topic	**# of 75-min. lect. (2/wk)**	**# of 50-min. lect. (3/wk)**
Ch. 1	1	1
Ch. 2	1.5	2
Ch. 3	1	2
Ch. 4	1.5	3
Test #1	1	1
Ch. 5	2	3
Ch. 6	2	3
Ch. 7	1	2
Ch. 8	2	3
Test #2	1	1
Ch. 9	2	3
Ch. 10	1.5	2
Ch. 11	1.5	2
Ch. 12	1	2
TOTAL	**20**	**30**

SCHEDULE FOR A THREE-QUARTER COURSE: SECOND QUARTER Waves, Electromagnetism		
Topic	**# of 75-min. lect. (2/wk)**	**# of 50-min. lect. (3/wk)**
Ch. 13	2	3
Ch. 14	2	3
Test #1	1	1
Ch. 19	2	3
Ch. 20	2	3
Ch. 21	3	5
Test #2	1	1
Ch. 22	2	3
Ch. 23	3	5
Ch. 24	2	3
TOTAL	**20**	**30**

SCHEDULE FOR A THREE-QUARTER COURSE: THIRD QUARTER Fluids, Thermal Physics, Light and Optics, Modern Physics		
Topic	**# of 75-min. lect. (2/wk)**	**# of 50-min. lect. (3/wk)**
Ch. 15	2	3
Ch. 16	1	2
Ch. 17	1.5	2
Ch. 18	1.5	3
Test #1	1	1
Ch. 25	1.5	2
Ch. 26	2.5	3
Ch. 27	1	2
Ch. 28	2	3
Test #2	1	1
Ch. 29	1.5	2
Ch. 30	1.5	3
Ch. 31	1	2
Ch. 32	1	1
TOTAL	**20**	**30**

Sample Syllabus

The following is a syllabus from a first-semester physics class. It contains information about class policies, test dates, grading, a list of topics covered by date and the associated reading assignments, and student strategies for success. Homework assignments with due dates can be included in the syllabus or on a separate handout. If your course has a laboratory component, you will also need to provide students with a lab schedule and guidelines for laboratory write-ups.

PHYSICS 131 SYLLABUS

Instructor: Judy Beck
Office: RBH 121A E-mail: jbeck@unca.edu
Office hours: MWF: 9:00-10:00 am, TR 3:00-4:00 pm, and by appointment, or feel free to just drop in!

Welcome to Introductory Physics I! In this semester, we will discuss motion, energy, momentum, fluids, waves, and thermodynamics. We will cover most of the first 18 chapters (volume 1) of *Physics* by James Walker. Please read over the policies and guidelines below, and see me if you have any questions or concerns.

Class Attendance: Attendance and participation in classroom activities are extremely important. Bring your calculator to class every day. It is also important to be on time for class. Handouts are distributed and announcements are made at the beginning of class. You are responsible for all information and instructions discussed in class whether or not you were present.

Prerequisite: MATH 164 or equivalent background in algebra, trigonometry, and pre-calculus is required. Do not take PHYS 131 at this time if you do not have this minimum mathematics prerequisite.

Laboratory: Laboratory attendance is required. One absence may be excused, but a second absence will lower your final grade for the course by one full level. Lab manuals should be purchased from the bookstore before the first lab meeting. You must also supply graph paper. Lab reports are to be completed and turned in by the end of the lab period unless otherwise noted.

Homework and Quizzes: Homework is due at the beginning of class on the dates indicated on the assignment sheet. Late homework will not be accepted. Solutions to all homework problems, practice problems, quizzes, and tests will be placed on reserve in the library. Discussion of homework problems is allowed and encouraged; however, copying of homework is not. All work submitted should represent your own best effort. Announced quizzes consisting of problems and conceptual questions will occasionally be given. Reading quizzes, taken on the Web, are due prior to class each Thursday.

Tests: Three tests will be given during the semester on the following dates: Tuesday, February 19; Thursday, March 28; Tuesday, April 25. Each test will include material covered in class up to that point, with emphasis on material covered since the last test. The final exam will be comprehensive and is scheduled for Tuesday, May 14, at 9:25 a.m. Please be sure to bring your calculator to all tests. Calculators may not be shared.

Grade Composition: The final grade will be determined by the following weighting:

Tests (3 @ 18% each):	54%
Final Exam:	25%
Homework & Quizzes:	15%
Lab reports:	6%

Grading Scale: The following scale will be used for letter grades:

A: 90–100 B: 80–89 C: 70–79 D: 60–69 F: less than 60

Academic Dishonesty: Students are expected to perform their own work on all assignments in this course. Dishonesty on an exam, quiz, homework, or lab report will result in a grade of zero for that assignment. Severe cases will result in a failing grade for the course.

<u>Schedule of topics and text assignments</u>: In order to prepare for class, please complete each reading assignment before the class during which the topic is discussed, and take the WebCT reading quiz for each chapter.

Week	Dates	Topics	Reading Assignment
1	Jan. 17	introduction; math review	Ch 1: all
2	Jan. 22, 24	kinematics: speed, velocity, acceleration, free fall	Ch 2: all
3	Jan. 29, 31	2-D motion: vectors, relative velocity, projectile motion	Ch 3: all
4	Feb. 5, 7	more on projectiles; force, inertia, Newton's 1st law	Ch 3: section 4 Ch 4: sections 1–2
5	Feb. 12, 14	Newton's 2nd and 3rd laws; friction	Ch 4: sections 3–6
6	Feb. 19	**Test #1: Chapters 1–4**	
	Feb. 21	energy, work, kinetic energy	Ch 5: sections 1–3
7	Feb. 26, 28	potential energy, conservation of energy, power; linear momentum and impulse	Ch 5: sections 4–6 Ch 6: sections 1–2
8	Mar. 5, 7	conservation of momentum, collisions, center of mass; circular motion	Ch 6: sections 3–5 Ch 7: sections 1–4
9	Mar. 12, 14	**Spring Break!**	
10	Mar. 19, 21	gravitation; rotational motion, torque, equilibrium	Ch 7: sections 5–6 Ch 8: sections 1–2
11	Mar. 26	rotational motion, continued	Ch 8: sections 3–5
	Mar. 28	**Test #2: Chapters 5–8**	
12	Apr. 2, 4	solids and fluids	Ch 9: all
13	Apr. 9, 11	temperature; heat, specific heat	Ch 10: all Ch 11: sections 1–2
14	Apr. 16, 18	heat, phase changes, heat transfer; thermodynamics	Ch 11: sections 3–4 Ch 12: sections 1–2
15	Apr. 23	thermodynamics, continued	Ch 12: sections 3–4
	Apr. 25	**Test #3: Chapters 9–12**	
16	Apr. 30, May 2	vibrations and waves	Ch 13: all
17	May 7, 9	sound	Ch 14: all
18	May 14	**Final Exam**	

Hints for best performance:

- Prepare for class; read material in the text *before* the lecture. Then read the material again after class discussions of the topics.
- Use your resources, including the information and hints on WebCT.
- Don't miss class; get notes from someone if you have an unavoidable absence.
- Study.
- Review and practice math as necessary.
- Participate in class. Bring you calculator every day.
- Practice, practice, practice.
- Let me know how you're doing!

Specifically related to problem solving and homework:

- Make sure you understand what the problem is asking before you begin plugging numbers into equations.
- Try the worked examples in the book and from your notes—without peeking!
- Try problems that have solutions in the back of the book.
- Draw a picture whenever possible.
- Pay attention to units.
- Don't expect to always work straight through a problem. Wrong turns and dead ends are often instructive.
- Check results to make sure they are reasonable.
- Get help—but only after trying the problem yourself.

And finally:

Keep in mind that learning physics involves both understanding concepts and solving problems. The two processes are inseparable and support each other. For instance, you will probably find that you understand a concept somewhat, then work your way through a related problem, which illuminates the concept even more, making the next problem easier to solve, etc. Finally it will all come together and you will understand the concepts and can whiz through the problems!

ENJOY!

Teaching Resources

Supplementary materials

In addition to this *Instructor's Resource Manual*, there are a number of other supplements to *Physics, second edition,* by J. S. Walker available from the publisher. The *Instructor's Solutions Manual* (Vol. I: 0-13-101479-X; Vol. II: 0-13-141660-X) contains detailed, worked solutions to every problem in the text. The *Test Item File* (0-13-101480-3), revised by Edson Justiniano (East Carolina University) contains approximately 2800 multiple-choice, short answer, and true/false questions, many conceptual in nature, referenced to the corresponding text sections. The *Transparency Pack* (0-13-140649-3) has 400 full-color transparencies of images from the text, including many of those mentioned in this manual. Finally, the *Instructor Resource CD-ROM* (0-13-101481-1), prepared by Sue Willis (Northern Illinois University) contains all text illustrations, in various formats. The CD also contains TestGen EQ, an easy to use, fully networkable software program for creating tests ranging from short quizzes to long exams. This CD-ROM set also contains electronic versions of the Instructor's Resource Manual, Instructor's Solutions Manual, and the End of Chapter Problems.

Be sure to visit the *Companion Website* (http://www.prenhall.com/walkerphysics), which contains excellent resources for both instructors and students. Students can practice problem-solving skills in self-assessment modules, explore and refine their understanding of physics, and learn to connect physics to the world around them. All the materials on the Web site are in addition to and not duplicative of material in the text. The site grades and scores all objective questions, the results of which can be automatically e-mailed directly to a professor or TA if so desired. Also included is an on-line syllabus builder for professors. See the text Preface for more details.

Additional study and assessment activities, including *Physlet® Illustrations and Physlet® Problems,* are featured in the Course and Homework Management systems (BlackBoard, WebCT, CourseCompass, PHGradeAssist) for *Physics, second edition.* Navigate to http://cms.prenhall.com/ for more information.

Demonstration resources

What would physics be without demonstrations? They provide excellent illustrations of important concepts, and besides that, they're fun. Even when they don't work (and believe us, sometimes they won't!), something can be learned from them. We have found that demonstrations are most effective when they involve students as much as possible. In addition to having students actually help perform the demonstrations, engage the whole class by asking for predictions or explanations. We've included a few demonstrations in each of the chapter sections that follow, but we encourage you to also check the resources listed below, ask your colleagues for favorites, and experiment with ideas of your own.

Edge, R. D., *String and Sticky Tape Experiments*. College Park, MD: American Association of Physics Teachers, 1981.

Ehrlich, R., *Turning the World Inside Out and 174 Other Simple Physics Demonstrations*. Princeton, NJ: Princeton University Press, 1990.

Ehrlich, R., *Why Toast Lands Jelly-Side Down: Zen and the Art of Physics Demonstrations*. Princeton, NJ: Princeton University Press, 1997.

Freier, G. D. and F. J. Anderson, *A Demonstration Handbook for Physics*. College Park, MD: American Association of Physics Teachers, 1981.

Meiners, H., editor, *Physics Demonstration Experiments*. New York: The Ronald Press Company, 1970.

Interactive teaching resources

There are a number of excellent books available to help instructors create a more active and interactive learning environment for students. A few are mentioned here that include valuable suggestions about teaching practices as well as many specific activities, quizzes, warm-up questions, and exercises. (For more information on these resources, see http://www.prenhall.com/tiponline/.)

Just-In-Time Teaching: Blending Active Learning with Web Technology, by G. Novak, E. Patterson, A. Gavrin and W. Christian (Prentice Hall, 1999), contains a description of the *Just-in-Time Teaching* strategy designed to enhance student learning. The book also contains many content-specific Web assignments students perform outside of class. Chapters 8 and 9 consist of warm-ups and puzzles from mechanics, thermodynamics, electricity and magnetism, and optics. Warm-ups are designed to help pique students' interest and prepare them for class. Puzzles are more complex and are attempted by students after the topic has been covered in the course.

Peer Instruction: A User's Manual, by Eric Mazur (Prentice Hall, 1997), describes an interactive teaching style developed by the author. It includes two nationally recognized evaluation tools, the Force Concept Inventory and the Mechanics Baseline Test, which can be used as pretests and post-tests to evaluate teaching effectiveness and student learning. In addition, it contains reading quizzes, "Conceptests," and conceptual exam questions for all main topics covered in a one-year introductory physics course. All questions are also in ready-to-print format on the enclosed diskette.

Physlets: Teaching Physics with Interactive Curricular Material, by W. Christian and M. Belloni (Prentice Hall, 2001), introduces Java physics applets that are designed to deal with individual concepts in physics. Physlets are interactive Web animations, which can be used in a variety of applications. This book teaches readers how to author physlets and includes a CD with physlets from all major topics in introductory physics.

Ranking Task Exercises in Physics, by T. O'Kuma, D. Maloney, and C. Hieggelke (Prentice Hall, 2000), is a thorough collection of exercises that require students to rank variations of a particular physical situation on a specified basis. These exercises are quite useful in developing and testing conceptual understanding.

Tutorials in Introductory Physics, by L. McDermott, P. Shaffer, and the Physics Education Group (Prentice Hall, 2002), contains supplementary instructional materials for a standard introductory physics course. The emphasis in the tutorials is on the development of important physical concepts and scientific reasoning skills rather than quantitative problem solving. The tutorials include pretests, worksheets, and homework assignments for topics from mechanics, electricity and magnetism, waves, and optics.

Physics education resources

The area of physics education research has produced lively discussions and resulted in numerous contributions to the literature in recent years. The following articles about teaching physics represent a sample that should prove interesting to both new and veteran instructors.

Appelquist, T. and Shapero, D., "Physics in a New Era," *Physics Today* (November 2001), pp. 34–39. Contains recommendations for strengthening physics in light of the needs of a rapidly changing society.

Bianchini, J., Whitney, D., Breton, T. and Hilton-Brown, B., "Toward Inclusive Science Education: University Scientists' Views of Students, Instructional Practices, and the Nature of Science," *Science Education* (January 2002), pp. 42–78. An excellent article with recommendations on ways to improve undergraduate science education.

Crouch, C. and Mazur, E., "Peer Instruction: Ten Years of Experience and Results," *American Journal of Physics* (September 2001), pp. 970–977. A description of the Peer Instruction teaching technique and an analysis of its success as compared to traditional lecture methods of teaching introductory physics.

Ehrlich, R., "How Do We Know if We Are Doing a Good Job in Physics Teaching?" *American Journal of Physics* (January 2002), pp. 24–29. An article based on a talk delivered by the author upon receiving the 2001 AAPT Award for Excellence in Undergraduate Teaching.

Fagen, A., Crouch, C. and Mazur, E., "Peer Instruction: Results from a Range of Classrooms," *The Physics Teacher* (April 2002), pp. 206–209. Reports student gains in the Force Concept Inventory score after participation in classes using the Peer Instruction pedagogical method.

Greca, I. and Moreira, M., "Mental, Physical, and Mathematical Models in the Teaching and Learning of Physics," *Science Education* (January 2002), pp. 106–121. An excellent article on the value of mental modeling in the teaching of physics.

Kim, E. and Pak, S.-J., "Students Do Not Overcome Conceptual Difficulties After Solving 1000 Traditional Problems," *American Journal of Physics* (July 2002), pp. 759–765. Investigates the relationship between traditional problem solving and conceptual understanding.

Lindenfeld, P., "Format and Content in Introductory Physics," *American Journal of Physics* (January 2002), pp. 12–13. A thoughtful guest comment on how little introductory physics courses have changed compared to the great changes in the field of physics.

Lopez, R. and Schultz, T., "Two Revolutions in K-8 Science Education," *Physics Today* (September 2001), pp. 44–49. Discusses two major changes taking place in precollege science education and the role of the scientific community in affecting change. Particularly interesting for students who are planning on teaching.

May, D., and Etkina, E., "College Physics Students' Epistemological Self-Reflection and Its Relationship to Conceptual Learning," *American Journal of Physics* (December 2002), pp. 1249–1258. An interesting article on students' self-reflection and its use to improve conceptual learning in physics classes.

Meltzer, D., "The Relationship Between Mathematical Preparation and Conceptual Learning Gains in Physics: A Possible 'Hidden Variable' in Diagnostic Pretest Scores," *American Journal of Physics* (December 2002), pp. 1259–1268. Suggests that students' preinstructional mathematical skills may be a significant factor in learning gains during physics instruction.

Meltzer, D. and Manivannan, K., "Transforming the Lecture-Hall Environment: The Fully Interactive Physics Lecture," *American Journal of Physics* (June 2002), pp. 639–654. Description of the development, application, and evaluation of active-learning techniques to large lecture classes.

Seymour, E., "Tracking the Processes of Change in US Undergraduate Education in Science, Mathematics, Engineering, and Technology," *Science Education* (January 2002), pp. 79–105. A history of efforts to improve science, math, engineering, and technology education.

Steinberg, R. and Donnelly, K., "PER-Based Reform at a Multicultural Institution," *The Physics Teacher* (February 2002), pp. 108–114. Describes reforms made in physics education at the City College of New York, based on physics education research.

Van Domelen, D., and Van Heuvelen, A., "The Effects of a Concept-Construction Lab Course on FCI Performance," *American Journal of Physics* (July 2002), pp. 779–780. An interesting report on the effects of two different lab curricula on student learning in introductory physics.

See also *Physics Education Research: A Supplement to the American Journal of Physics*, Supplement 1 to vol. 68, No. 7, (July 2000). Contains many wonderful articles on physics education.

Physics organizations

The following organizations are concerned about issues in the teaching of physics and have helpful resources available.

American Association of Physics Teachers (AAPT). Publishes the *American Journal of Physics* and *The Physics Teacher*. Sponsors workshops and meetings on teaching innovations. (http://www.aapt.org)

American Institute of Physics (AIP). Publishes *Physics Today.* Excellent educational materials. (http://www.aip.org)

American Physical Society (APS). Publishes the *Physical Review* journals. (http://www.aps.org)

Astronomical Society of the Pacific (ASP). Excellent educational materials related to astronomy. (http://www.aspsky.org)

Materials and equipment

The following companies are good sources of materials and equipment for demonstrations and labs.
Edmund Scientific (http://www.edsci.com). Educational science products can be found at http://www.scientificonline.com.

Fisher Scientific (http://www.fishersci.com/main.jsp). Fisher Science Education (http://www.fisheredu.com).

Frey Scientific (http://www.freyscientific.com).

Klinger Educational Products Corporation (http://www.KlingerEducational.com).

Ohaus Corporation (http://www.ohaus.com).

PASCO (http://www.pasco.com).

Sargent-Welch (http://www.sargentwelch.com). Also contains Cenco Physics product listings.

Organization of the IRM

We have organized the rest of this manual by textbook chapters. In each section you will find a chapter outline, summary, list of major concepts, and teaching suggestions and demonstrations. At the end of each section we have listed the transparencies and Physlet® illustrations available for the chapter. In addition, we suggest readings pertaining to the chapter content and materials that may be useful for demonstrations. We conclude each section with space for your notes.

A Final Word

We find the teaching of introductory physics extremely stimulating and rewarding, and hope you do as well. One of our goals is to change—and indeed enrich—the way our students look at the world. After taking physics, one has a new perspective on the world, whether admiring a rainbow or riding a bicycle. Although the specifics of formulas will eventually be forgotten, we hope that our students retain a sense of the relationships in physics and an awe for the workings of the universe based on knowledge and understanding. Students can always go look up equations, but the skills to think critically, solve problems, and evaluate issues from an educated perspective will serve them well no matter what their future endeavors may be.

We would like to thank Christian Botting of Prentice Hall for his guidance and support during this project. A special thank-you goes to Jerome, Caitlin, and Duncan Hay, and Jac, Katie, and Michael Whatley for their encouragement and understanding throughout the writing process. We are indebted to our mentors, colleagues, and students who inspire our love of physics and our teaching pursuits.

Katherine M. Whatley
Judith A. Beck
Department of Physics
The University of North Carolina at Asheville
whatley@unca.edu
jbeck@unca.edu
March 2003

Chapter 1: Introduction

Outline

1-1 Physics and the Laws of Nature
1-2 Units of Length, Mass, and Time
1-3 Dimensional Analysis
1-4 Significant Figures
1-5 Converting Units
1-6 Order-of-Magnitude Calculations
1-7 Problem Solving in Physics

Summary

Chapter 1 presents the general definition of *physics* and some of the conceptual tools needed to begin its study. The chapter includes discussion of units and conversion factors, dimensional analysis, significant figures, and scientific notation. In addition, the usefulness of order-of-magnitude calculations and tips on problem-solving methods in physics are addressed.

Major Concepts

By the end of the chapter, students should understand each of the following and be able to demonstrate their understanding in problem applications as well as in conceptual situations.

- General definition of physics
- Systems of units
    - Length
    - Mass
    - Time
- Dimensional analysis
- Significant figures
    - Addition and multiplication
    - Scientific notation
- Unit conversions
- Estimating (order of magnitude calculations)
- Problem-solving strategies

Teaching Suggestions and Demonstrations

Many of the students who take a college physics course will have seen most of the material presented in this chapter in previous courses. The general tendency is to skip over this chapter or have the students read it and not discuss it in class. We don't recommend this. While many students may have seen the material, most will not have internalized it. It is a good idea to spend a little time introducing these concepts. Talk about them again each time you work an example in class. Even good students seem to have trouble remembering to use significant figures correctly.

Section 1-1

Begin by **justifying the study of physics**. Many of the students will be in this course because it satisfies a requirement. You need to convince them that it will also be interesting (and fun!). Physics provides us with ways to model the universe. Models are useful in the understanding of phenomena and also in the prediction of future behavior. (You might point out how physics is different from what psychics claim to do!)

Sections 1-2 and 1-3

Most students are familiar with **definitions of units** and **systems of units**. They will be comfortable with British and metric units of length and time. Plan to spend a little time on the idea and unit of mass; most students will not have heard of the British unit (slug). This is a good time to emphasize the importance of units. (See Table 1-4 for a list of common prefixes.)

➲ **DEMO 1-1** Measure a convenient object (a table is often available and works well). Announce to the class that its length is "two" (or something close to its length in meters). Ask the class if this information conveys anything useful. Usually someone will say "Two what?," which gives you the perfect opening to talk about the importance of units.

In September 1999, a probe launched by NASA was lost in the atmosphere of Mars because the engineers who built the engines were working in British units and the scientists who were controlling the engines were working in SI units. (See Resource Information for reference.) This makes a good story and (again) lets you emphasize that units are part of every result.

Dimensional analysis will likely be a new concept for most students. Its usefulness as a problem-solving strategy will not be immediately obvious. Discuss it here and continue to point out the consistency of units every time a new formula is introduced. Work Exercise 1-2 in class.

Section 1-4

Most students will also have heard of the concept of **significant figures** but will not know the rules or have any idea of its importance. It is a good idea to go over the rules and to discuss how they follow common sense. For example, consider a highway engineer with a 100-ft tape measure. The tape is cut accidentally. When the engineer splices it back together, she has lost about half an inch near the beginning of the tape. Does that half inch matter? It depends. If she is using the tape to measure flower beds for a highway beautification program, then no, the lost half inch doesn't matter. If she is using the tape to measure replacement glass for a garage window, the half inch matters a great deal!

You will also need to talk to students about significant figures and calculators. Just because nine digits appear on the display does not mean there are nine significant figures!

➲ **DEMO 1-2** The 2000 U.S. Presidential election is another source for discussion material on significant figures, especially if you have political science or history majors in your class. Choose any state where the vote count was close (Florida, New Mexico, and Wisconsin are examples) and discuss the vote difference in terms of significant figures. (You can begin talking about error analysis with this example, too, but only at a very general level.)

It is sometimes easier to discuss significant figures in the context of **scientific notation**. The number of significant figures is often more clear when a number is written in scientific notation. Some students may not be comfortable with exponents; you will need to make sure that the class understands that 10 x 10 = 10^2, etc.

Sections 1-5 through 1-7

Students generally catch on to basic **unit conversion** rather quickly, but they have trouble with converting compound units (miles/hour to m/s, for example) and powers of units (cm^2 to m^2). Go over these conversions now, and be prepared to go over them often.

➲ **DEMO 1-3** Make a cube measuring 10 cm on a side. Cardboard or Styrofoam work well and are light. Mark off the centimeter lines. You can use this to illustrate the relationship between 1 cm and 1 m, between 1 cm^2 and 1 m^2, and between 1 cm^3 and 1 m^3. This is particularly effective if you also have a cube that is 1 m on a side. (A 1-m cube is much harder to store.)

Emphasize the usefulness of **order-of-magnitude calculations**. You can estimate the amount of paint you need to paint a room, the amount of water needed to wash a car, the amount of food needed for a cookout – anything that catches the students' attention. Then discuss how order-of-magnitude calculations can help check for "reasonableness" of problem solutions.

➲ **DEMO 1-4** You can do a quick estimate of the number of times your heart beats in a year. Take your pulse for 6 seconds. Multiply by 10 and round off to a reasonable number. That gives the number of times your heart beats a minute. Then use rounded-off unit conversions to estimate the number of beats in a year. (Number of beats in a minute x 60 minutes/hour x 20 hours/day x 400 days/year.)

The end of this chapter has a nice section on problem solving in physics. You can talk about this in general, but students understand problem-solving techniques best in the context of solving a problem. Students generally like a little routine and a little predictability in class. Write down a few key rules (keep them short) and repeat them mantra-like every time you solve a problem. (What is given? Draw a picture. Think. What are the equations? Solve. Check your answer. UNITS!) Soon the students will repeat them as well, both in class as you are working examples, and out of class when they are working on their own.

Resource Information

Transparencies

1. Table 1-1	Typical Distances
2. Table 1-2	Typical Masses
3. Table 1-3	Typical Times
4. Table 1-4	Common Prefixes

Physlet® Illustrations

1-1 Introduction to Physlet Illustrations

Suggested Readings

Bergquist, J., Jefferts, S. and Wineland, D., "Time Measurement at the Millennium," *Physics Today* (March 2001), pp. 37-42. One of two articles celebrating the centennial of the National Institute of Standards and Technology (NIST).

Frisch, H., "No Kid Should Know That Much About . . .," *Physics Today* (October 1999), pp. 71-72. An opinion piece on science education. Mentions several great demonstrations.

Goodwin, I., "Washington Briefings: One Too Many Mishaps on Voyages to Mars," *Physics Today* (January 2000), p. 47. A short summary of the problems with recent NASA Mars missions.

Haseltine, E., "The Greatest Unanswered Questions of Physics," *Discover* (February 2002), pp. 36-42. An interesting general article on current issues in physics.

Hillger, D., "Metric Units and Postage Stamps," *The Physics Teacher* (November 1999), pp. 507-510. Great article on SI units.

Keeports, D., "Addressing Physical Intuition—A First-Day Event," *The Physics Teacher* (May 2000), pp. 318-319. An example of what to do on the first day of class.

Mohr, P. and Taylor, B., "Adjusting the Values of the Fundamental Constants," *Physics Today* (March 2001), pp. 29-34. One of two articles celebrating the centennial of the National Institute of Standards and Technology (NIST).

Romer, R., "Units—SI-Only, or Multicultural Diversity?," *American Journal of Physics* (January 1999), pp. 13-16. An editorial discussing choice of units in physics. See also numerous letters to the editor in response in the June 1999 issue.

Wheeler, D. and Mazur, E., "The Great Thermometer Challenge," *The Physics Teacher* (April 2000), p. 235. An interesting activity that emphasizes the importance of critical thinking.

Notes and Ideas

*Class time spent on material: Estimated:*______ *Actual:*_____________

Related laboratory activities:

Demonstration materials:

Notes for next time:

Chapter 2: One-Dimensional Kinematics

Outline

Summary

After the groundwork is laid in Chapter 1, Chapter 2 begins the real study of physics. One-dimensional kinematics is the study of motion in a straight line without regard to its causes. Many of the concepts in this chapter, such as velocity and acceleration, are familiar to students from everyday experiences, like driving a car. However, most students will not know the physics definitions of these same terms or how they relate to one another. In addition, common misconceptions due to the presence of air resistance and friction in the real world need to be addressed. Chapter 2 lays the foundation for two-dimensional motion addressed in later chapters.

Major Concepts

By the end of the chapter, students should understand each of the following and be able to demonstrate their understanding in problem applications as well as in conceptual situations.

- Position, distance and displacement
- Speed and velocity
  - Average
  - Instantaneous
  - Constant
- Acceleration
  - Average
  - Instantaneous
  - Constant
- Graphs of position versus time, velocity versus time, and acceleration versus time
- Equations of motion with constant acceleration
- Free fall

Teaching Suggestions and Demonstrations

Since problem solving is a large component of most physics courses, it is a good idea to emphasize it from the beginning. A short quiz at the end of the chapter can be helpful at this early stage to give students practice in solving problems on their own.

Sections 2-1 through 2-4

Beginning physics students already know the difference between distance and speed. However, you will need to help them differentiate between **distance** and **displacement**, and between **speed** and **velocity**. Also discuss constant, average, and instantaneous speed and velocity. Conceptual Checkpoint 2-1 is an excellent one for students to try since the result may surprise them.

The **units of acceleration** are initially puzzling. Have students calculate an acceleration they are familiar with, such as the acceleration of a car on the entrance ramp to the highway. They can initially use units that are intuitively sensible, such as miles per hour per second, and then convert to the more conventional m/s^2 to get a feel for the units they will be using in physics class.

Students often have trouble understanding the important differences between **constant velocity** and **constant acceleration**. A helpful approach is to directly compare these two cases. Imagine, for instance, a cart with a spark timer that leaves marks on a straight track at one-second intervals. Students can determine distance traveled by the cart during each second when it is moving with a constant velocity and then again when it is moving with a constant acceleration. When the distances are plotted on a one-dimensional line, it is apparent that the cart covers equal distances in each one-second interval for the constant velocity case but increasingly larger distances in each successive second for the constant acceleration case.

Graphs can also often be difficult for beginning physics students to interpret. The exercise above, in which the position of a car each second is plotted on a straight line, is a nice intermediate step leading into actual position-versus-time graphs, as illustrated by comparing Figures 2-3 and 2-4. Next, create the corresponding velocity- and acceleration-versus-time graphs and determine the significance of slopes and areas under the curves. Figures 2-5 through 2-7 help students visualize velocity from a position-versus-time graph. Relating the three graphs of position, velocity, and acceleration versus time to each other and to the description of the motion requires practice and is a very valuable exercise, whether used quantitatively or conceptually.

➲ **DEMO 2-1** A motion sensor with computer interface and software to plot the graphs of actual moving objects is an excellent demonstration tool for graphical understanding and interpretation. (See Resource Information.) One approach is to have students sketch their predictions for a certain motion and then receive immediate feedback from the computer. Alternatively, students can be shown a graph on the computer and try to move an object or themselves in front of the motion sensor to duplicate it.

Throughout this chapter, the significance of **positive and negative signs** needs to be emphasized. The meanings of positive and negative displacement and velocity are fairly straightforward. Acceleration is more confusing. Some numerical examples can help convince students that a negative acceleration does not necessarily mean an object is slowing down. (An object slowing down while traveling in the positive direction and an object speeding up while traveling in the negative direction both have negative accelerations.)

➲ **DEMO 2-2** The simple demonstration of dropping objects simultaneously from the same height helps convince students that **acceleration due to gravity** really is the same for all objects. Choosing a wide variety of objects can also help them get a feel for the conditions under which air resistance can and cannot be ignored.

➲ **DEMO 2-3** One very simple but effective demonstration is to drop a quarter or a ball and a piece of notebook paper simultaneously from the same height, as illustrated in Figure 2-17. First give the class a chance to guess what they think will happen. They have just learned that the acceleration due to gravity is the same for all objects, but they will usually revert to their experience and guess that the quarter will fall first. Then drop the quarter and paper. Discuss with the class why the quarter fell first. Someone will mention air resistance. Discuss ways in which air resistance could be taken out of the problem. Crush the paper into a ball and again poll the class about which object will hit the ground first. This time about half the class will guess that the two objects will fall at the same rate. Drop the quarter and paper and discuss the results. Students respond well to this demonstration; it gives them dramatic evidence of the role of air resistance in free fall and information about when it can be ignored.

➲ **DEMO 2-4** Another very simple demonstration involves filling a Styrofoam (or paper) cup with water and punching a hole with a pencil in the side near the bottom. (It is best to do this over a sink or a trash can, since it can be messy.) When the cup is held stationary, water shoots out of the hole. (You can hold a nice discussion about the shape of the curve the water makes if the cup is large enough to hold sufficient water.) When the cup is dropped, water will not come out of the hole, since the cup and the water are falling at the same rate. Be sure to ask the class what they think will happen before dropping the cup; often no one will guess that no water will emerge from the hole.

Sections 2-5 through 2-7

Free fall provides a wonderful opportunity to emphasize major points regarding position, velocity and acceleration for the case of constant acceleration. Although most students understand intuitively that the velocity of a ball thrown up in the air is zero at the top of its trajectory, many assume incorrectly that the acceleration there is zero as well. Graphs of y versus t, v versus t, and a versus t are very useful in finding and correcting student misconceptions about acceleration for objects in free fall. Carefully go over the graphs in Figure 2-20 for a lava bomb projected straight upward. Emphasize the fact that acceleration is a rate of change. Even though the object moves first up and then down, the rate of change of its velocity is a constant $-9.81\ \text{m/s}^2$.

➲ **DEMO 2-5** A free-fall apparatus with spark timer can demonstrate change in position for objects undergoing acceleration due to gravity. The sparks mark the paper at equal time intervals, and therefore the distance between adjacent sparks increases as the plummet accelerates. This demonstration is useful for helping students visualize constant acceleration and can also be used to calculate the acceleration due to gravity. (See Resource Information.)

Finally, a solid understanding of the **equations of motion for constant acceleration,** Equations 2-7, 2-10, 2-11, and 2-12, provides students with an excellent base upon which to build in Chapter 4 when two-dimensional motion is considered. Initially, students may need guidance in using the equations and in recognizing which terms are constants and which are variables. The importance of consistency with sign conventions also needs to be emphasized, as can be demonstrated by comparing a problem in which an object is thrown straight up from a bridge to one in which an object is thrown straight down from the same bridge at the same speed. Although the speeds are the same, the velocities have different signs and so are not the same, resulting in very different outcomes. Explicitly choosing a positive direction at the outset of each problem reminds students of this important distinction. Try Conceptual Checkpoint 2-5 with students for another example of objects in free fall.

➲ DEMO 2-6 Students get a kick out of measuring their own reaction time and comparing it to others. Have students work in pairs with one holding a ruler vertically so that the zero mark is level with the second student's fingertips. The first student lets go, and the second catches the ruler. The time it took to catch the ruler is calculated using g and the distance the ruler fell, read directly from where the catcher's fingers snagged the ruler. As an added bonus, make a bar graph of the reaction times from the whole class and use the graph to introduce bell curves and standard deviations.

Resource Information

Transparencies

5.	Figure 2-2	One-dimensional coordinates
6.	Figure 2-3	One-dimensional motion along the x axis
	Figure 2-4	Motion along the x axis represented with an x-versus-t graph
7.	Figure 2-7	Instantaneous velocity
	Table 2-1	x-versus-t values for Figure 2-7
8.	Figure 2-8	Graphical interpretation of average and instantaneous velocity
9.	Figure 2-9	v-versus-t plots for motion with constant acceleration
10.	Figure 2-10	Graphical interpretation of average and instantaneous acceleration
11.	Figure 2-11	Cars accelerating or decelerating
	Figure 2-12	v-versus-t plots with constant acceleration
12.	Figure 2-13	The average velocity
13.	Example 2-9	Catching a Speeder
14.	Figure 2-18	Free fall from rest

Physlet® Illustrations

2-1 Determination of Average Speed
2-2 Velocity and Speed
2-3 Graphical Determination of Velocity
2-4 Graphical Determination of Acceleration
2-5 Free Fall
2-6 Air Resistance

Suggested Readings

Andereck, B., "Measurement of Air Resistance on an Air Track," *American Journal of Physics* (June 1999), pp. 528-533. Excellent experiment for a more in-depth examination of air resistance.

Conderle, L., "Extending the Analysis of One-Dimensional Motion," *The Physics Teacher* (November 1999), pp. 486-489. Thorough treatment of displacement, velocity, and acceleration-versus-time graphs for one-dimensional motion.

Levi, B., "Atom Interferometer Measures g with Same Accuracy as Optical Devices," *Physics Today* (November 1999), p. 20. A description of an extremely accurate measurement of g.

McClelland, J., "g-whizz," *The Physics Teacher* (March 2000), p. 150. A discussion of the constant g.

Moreland, P., "Improving Precision and Accuracy in the *g* Lab," *The Physics Teacher* (September 2000), pp. 367-369. Discussion of a lab to determine acceleration due to gravity.

Patterson, J., "Physical Principles versus Mathematical Rigor," *The Physics Teacher* (April 2000), p. 214. Using a 1-D kinematics problem to illustrate how physics principles can simplify calculations.

Rist, C., "The Physics of . . . Baseballs: Foul Ball?" *Discover* (May 2001), pp. 26-27. An interesting article on how the construction of a baseball affects its performance.

Singh, K., "The Flight of the Bagel," *The Physics Teacher* (October 2000), pp. 432-433. Using a bagel and camcorder to determine the value of *g*.

Wick, K. and Ruddick, K., "An Accurate Measurement of *g* Using Falling Balls," *American Journal of Physics* (November 1999), pp. 962-965. A description of an experiment to determine the acceleration due to gravity, *g*, with an accuracy of about 1 part in 10^4.

Materials

The Pasco *Science Workshop*® is a good system for use in computer-based labs. Many bundles that include the Motion Sensor II (CI-6742) are available.

One free-fall apparatus is the Sargent-Welch/Cenco/Physics Behr Free-Fall Apparatus, model number CP00749-05. There are several required accessories: a six-volt AC/DC power supply (CP33031-00), a spark timer (CP31755-01) and wax-coated recording tape (WLS-65250-C).

Notes and Ideas

*Class time spent on material: Estimated:*______ *Actual:*____________

Related laboratory activities:

Demonstration materials:

Notes for next time:

Chapter 3: Vectors in Physics

Outline

3-1 Scalars versus Vectors
3-2 The Components of a Vector
3-3 Adding and Subtracting Vectors
3-4 Unit Vectors
3-5 Position, Displacement, Velocity, and Acceleration Vectors
3-6 Relative Motion

Summary

This chapter moves the concepts of position, displacement, velocity, and acceleration into the two-dimensional world. In two dimensions, direction can no longer be indicated simply by positive and negative signs. Vectors and vector manipulations, important for the remainder of the course, are introduced.

Major Concepts

By the end of the chapter, students should understand each of the following and be able to demonstrate their understanding in problem applications as well as in conceptual situations.

- Scalars (magnitude only)
- Vectors (magnitude and direction)
  - Components
  - Addition and subtraction
- Unit vectors
- Vector position, displacement, velocity, and acceleration
- Relative motion

Teaching Suggestions and Demonstrations

Understanding the difference between scalar and vector quantities is key to understanding many of the concepts in an introductory physics course. Vector manipulation is also an important skill for the students to master. Many of the students will not have seen vectors before. Spending class time going over scalars, vectors, and vector manipulations *now* will pay off later in the course.

Sections 3-1 through 3-4

Students usually understand the concept of **scalar** quantities. Ask the students to give examples of scalars, but be prepared to supply a few yourself (*e.g.*, mass, time, speed).

The concept of **vector** quantities is more problematic. Begin with the definition of a vector quantity (magnitude and direction) and decide how you will indicate these quantities for class. Books usually use boldface type, but it is difficult to do boldface on the board or on an overhead. An arrow over the vector quantity works well.

Position and displacement vectors are the simplest to describe. It is relatively easy for students to see that a position vector has both a magnitude and a direction.

➲ **DEMO 3-1** If you are in a room with a tile floor, you can use the tile as a grid and have students stand at the "head" and "tail" of a position vector. (If there is no tile pattern in your room, you can copy a piece of graph paper onto a transparency and use an overhead projector.) Have a student walk through several displacements and see how the position vector, measured from the origin, changes. This is also a good time to define the negative of a vector, a concept needed for vector subtraction.

Before tackling vector **components**, you will need to review the definitions of the sine, cosine, and tangent of an angle, the relationship between degrees and radians, and the Pythagorean theorem. (There is usually a wide range of understanding of these concepts among students enrolled in introductory physics classes.)

As you define the **scalar components of a vector**, be sure the students understand that the components are scalars and can be positive or negative. (See Figure 3-6.) Demonstrate that the trigonometric function associated with a particular component of a vector depends on the *angle* chosen. (The *x*-component is not always associated with the cosine function, and the *y*-component is not always associated with the sine function. Many students have difficulty with this idea.)

Begin treatment of the rules for **vector addition and subtraction** with the graphical case. You can use the tile floor or the graph paper on an overhead projector mentioned previously. When the students understand that vector addition is different from scalar addition, move on to the component method of vector addition and subtraction. (See Figures 3-11, 3-12, and 3-13.)

It's nice to remind students that they can use the graphical method to check their work with addition and subtraction of vectors by the component method. They can do a quick sketch of a problem and estimate the answer, then check to make sure their careful calculation makes sense.

A discussion of **unit vectors** follows logically from the component method. Once the vector is broken down into its components, it makes sense to leave it in component form. Future calculations will be easier. Define the unit vectors in the *x* and *y* directions and point out that unit vectors are dimensionless. The magnitude of the vector carries its units. Be sure to show several examples of vector addition and subtraction using unit vectors. Emphasize how easy it is to manipulate the vectors once they are written in unit vector form.

Sections 3-5 and 3-6

Now that the students are familiar with vectors in general and with position and displacement vectors, introduce **velocity and acceleration vectors**. Point out the vector nature of both quantities.

➲ **DEMO 3-2** Have the students think about velocity and acceleration in a car. If they travel around a curve at constant speed, does the velocity change? (Yes, because the direction of the velocity vector changes.) How many "accelerators" are there in a car? (Three—the accelerator pedal and the brake, which both change the speed and therefore change the velocity, and the steering wheel, which changes the direction of the velocity vector.) (See Figure 3-23.)

Go over Active Example 3-2 in class.

A treatment of **relative motion** involves the addition of velocity vectors. This will be the first time the students have tried to add and subtract vectors other than position vectors. Assure them that the rules for vector addition and subtraction are the same no matter what kind of vector is under consideration. Go over Figure 3-26 and work Example 3-2 with the students.

Resource Information

Transparencies

15. Figure 3-4	A vector and its scalar components
16. Figure 3-6	Examples of vectors with components of different signs
17. Figure 3-12	Graphical addition of vectors
18. Figure 3-13	Component addition of vectors
19. Figure 3-20	Average velocity vector
Figure 3-21	Instantaneous velocity vector
20. Figure 3-22	Average acceleration vector
21. Figure 3-23	Average acceleration for a car traveling in a circle with constant speed
22. Figure 3-24	Velocity and acceleration vectors for a particle moving along a winding path
23. Figure 3-27	Relative velocity in two dimensions
24. Example 3-2	Crossing a River

Physlet® Illustrations

3-1 Components of Vectors
3-2 Relative Motion

Suggested Readings

Black, H., "Vector Toy," *The Physics Teacher* (September 1998), p. 375. Describes the use of "walking toys" to demonstrate vector forces.

Durkin, T. and Graf, E., "Quibbles, Misunderstandings, and Egregious Mistakes," *The Physics Teacher* (May 1999), pp. 297-306. Discusses common mistakes in high school textbooks, with a nice section on vectors.

Larson, R., "Measuring Displacement Vectors with the GPS," *The Physics Teacher* (March 1998), p. 161. A nice use of GPS to talk about vectors.

Sheets, H., "Communicating with Vectors," *The Physics Teacher* (December 1998), pp. 520-521. A good exercise for teaching vectors.

Taylor, D., "Vector Video," *The Physics Teacher* (January 2001), p. 14. A great illustration of vector addition and relative motion.

Wheeler, D. and Charoenkul, N., "Whole Vectors," *The Physics Teacher* (May 1998), p. 274. A nice treatment of vectors and vector diagrams.

Widmark, S., "Vector Treasure Hunt," *The Physics Teacher* (May 1998), p. 319. A fun activity for teaching vectors.

Notes and Ideas

Class time spent on material: Estimated:_______ Actual:_______________

Related laboratory activities:

Demonstration materials:

Notes for next time:

Chapter 4: Two-Dimensional Kinematics

Outline

Summary

Chapter 4 extends the discussion of kinematics to two dimensions by building on students' knowledge of one-dimensional motion (Chapter 2) and vectors (Chapter 3). Projectile motion, or motion of an object under the influence of gravity only, is treated thoroughly in this chapter.

Major Concepts

By the end of the chapter, students should understand each of the following and be able to demonstrate their understanding in problem applications as well as in conceptual situations.

- Motion in two dimensions
  - Components of velocity and acceleration
  - Equations of motion for constant acceleration and constant velocity
- Projectile Motion
  - Acceleration due to gravity: *g*
  - Independence of horizontal and vertical motions
  - Air resistance
  - Basic equations
    - Special case: zero launch angle
    - General case
  - Characteristics of projectile motion

Teaching Suggestions and Demonstrations

The two-dimensional equations of motion may at first appear intimidating to students. Point out that they are really nothing more than applications of equations already introduced in the previous chapters. Encourage students to work many problems in this chapter to become comfortable with the variables and the initial conditions. Also remind them to use sketches to illustrate the problems for clarity.

Section 4-1

In order to derive the **equations of motion in two dimensions,** students must first understand that the horizontal and vertical motions are independent of each other. Acceleration in the *x*-direction will not affect velocity in the *y*-direction and vice versa. Once this concept is addressed, begin with the equations for one dimension and replace v, v_o, and a with v_x, v_{ox}, and a_x, pointing out that only the *x*-components of the velocity and acceleration are relevant to equations describing motion in the *x*-direction. The equations for the *y*-direction are identical, except that all *x*'s are replaced with *y*'s. After some examples, students should notice the importance of **time** in the equations. It is the only variable that does not have components and is also the only one shared by both x and y directions. Try Examples 4-1 and 4-2, or others like them, to illustrate both constant velocity and constant acceleration in two dimensions.

Sections 4-2 through 4-4

A **projectile** is any object that, after being launched into motion, is under the influence of gravity only. This is a broader definition than the common usage of the word, which conjures up images of cannon balls or food fights, so plenty of examples are helpful. A thrown baseball and a dropped rock are projectiles; a rocket firing its engines is not. As with free-fall motion (see Chapter 2), we assume that air resistance is negligible. In addition, the rotation and curvature of the Earth are ignored. Discuss the conditions under which these assumptions are valid.

The **basic equations of projectile motion,** Equation 4-6, are obtained directly from the two-dimensional equations of motion for constant acceleration with $a_x = 0$ and $a_y = -g$. The equations describing motion in the x-direction simplify to their constant velocity form. Also, the motion in the y-direction is independent of the x-velocity. For example, with the assumptions given above, a bullet dropped and a bullet fired horizontally from the same height over level ground will hit the ground at the same moment. Extend Conceptual Checkpoint 4-1 by asking which diver will hit the water first.

➲ **DEMO 4-1** To demonstrate the **independence of the *x* and *y* components of the velocity of a projectile**, have one student slide a quarter off a level table so that it leaves the table with an initial velocity only in the x-direction. A second student holds a second quarter at the edge of the table and, with it at eye level, attempts to drop it as the first quarter leaves the table. Students can predict which will hit the ground first and listen for the results. It should be apparent that they hit the ground at the same time. (If not, discuss reaction time and the difficulty of dropping the second quarter at exactly the right moment.) Since each has the same y-component of initial velocity, namely zero, they travel the vertical distance to the floor in the same amount of time. However, they do not land at the same place. Since their initial velocities have different x-components, their x-displacements are not identical.

➲ **DEMO 4-2** If you have or can construct the apparatus, the classic *Monkey and Hunter* (or *Shoot the Target*) demonstration is an excellent way to show the independence of the horizontal and vertical motions for projectiles as well as the fact that g is constant. (See Resource Information.) A hunter wants to shoot a monkey and isn't sure where to aim. He knows that as soon as the gun is fired, the monkey will be startled by the noise and let go of his branch and drop. So should the hunter fire directly at, above, or below the monkey? Most students will want to fire below, since they know that the monkey will fall while the bullet is on the way. However, the bullet is accelerated downward at the same rate, so in fact the hunter needs to aim directly at the monkey.

Emphasize the importance of **time** in the projectile motion equations. If a problem gives information about one direction, for example the initial height of a projectile, and asks for information about the other direction, for example the horizontal range, students must use the time variable to connect the two sets of equations. Example 4-6 is a good one to do with students; it uses the time variable to connect a ball moving in one dimension and a dolphin moving in two dimensions.

Section 4-5

The last section of the chapter contains further information about projectile motion. The **range** equation, Equation 4-12, is a specific application of the general equations for projectile motion. Before examining it, students can predict the angle of the initial velocity that will maximize the horizontal range of a projectile. Most will correctly predict 45°. See Figure 4-11 to compare the paths of projectiles with different launch angles.

➲ **DEMO 4-3** In order to test that the **maximum range of a projectile** is obtained with an initial velocity at an angle of 45°, try projecting a ball at various angles with a compressed spring. It is important to compress the spring the same amount each time, so that the initial speed is constant and the only quantity varying is the angle. (See Resource Information or build your own launcher!) If air resistance is indeed negligible, students will also discover that angles less than 45° and angles more than 45° by the same amount (such as 30 ° and 60 °) will result in equal ranges.

The concept of **symmetry in projectile motion** is extremely useful, and students can be encouraged to use it to their advantage. For example, a projectile launched over level ground will spend as much time on the trip up as on the trip down. If the projectile is launched from a cliff, the time it takes to reach its maximum height will be the same as the time it takes to fall from that height back to the level of the top of the cliff. Also, the projectile will be traveling the same speed when it is back to the same level, although the y-component of its velocity will now be down instead of up. See Conceptual Checkpoint 4-3 for an example.

Resource Information

Transparencies

25. Figure 4-1	Constant velocity
26. Table 4-1	Constant-Acceleration Equations of Motion
27. Figure 4-2	Acceleration in free fall
28. Figure 4-5	Trajectory of a projectile launched horizontally
29. Conceptual Checkpoint 4-1	
30. Figure 4-6	Projectile Motion with arbitrary launch angle
31. Figure 4-7	Snapshots of a trajectory
32. Example 4-6	A Leap of Faith
33. Figure 4-9	Projectiles with air resistance
34. Figure 4-10	Velocity vectors for a projectile launched at the origin
35. Figure 4-11	Range and launch angle in the absence of air resistance

Physlet® Illustrations

4-1 Motion in Two-Dimensions
4-2 Projectile Motion
4-3 The Monkey and Hunter
4-4 Range of a Projectile

Suggested Reading

Chow, J., Carlton, L., Ekkekakis, P. and Hay, J., "A Web-Based Video Digitizing System for the Study of Projectile Motion," *The Physics Teacher* (January 2000), pp. 37-40. A two-dimensional projectile motion lab using a camcorder for data collection.

Deakin, M. and Troup, G., "Approximate Trajectories for Projectile Motion with Air Resistance," *The American Journal of Physics* (January 1998), pp. 34-37. Describes approximations for the trajectories of projectiles under various laws of resistance.

Molina, M., "More on Projectile Motion," *The Physics Teacher* (February 2000), pp. 90-91. Discussion of the range of a projectile.

Price, R. and Romano, J., "Aim High and Go Far—Optimal Projectile Launch Angles Greater Than 45°," *The American Journal of Physics* (February 1998), pp. 109-113. An investigation of optimal projectile launch angles when air resistance is taken into account, with simple physical arguments that help explain the results.

Van den Berg, W. and Burbank, A., "Sliding Off a Roof: How Does the Landing Point Depend on the Steepness?" *The Physics Teacher* (February 2002), pp. 84-85. Examination of an object sliding off an elevated ramp, with a plot of horizontal projection versus angle of inclination.

Wetherhold, J., "A Toy Airplane for Projectile Motion Experiments," *The Physics Teacher* (February 2001), pp. 116-119. An inexpensive toy airplane modified to demonstrate various characteristics of projectile motion.

Materials

A "Projectile Launcher" can be obtained from Frey Scientific, catalog number S1900503.
Two variations of the "Shoot the Target" apparatus are available from Pasco, catalog numbers ME-6805 and ME-6826. (Projectile launchers from Pasco for use with the apparatus include ME-6801 and ME-6825).

Notes and Ideas

Class time spent on material: Estimated:______ Actual:_____________

Related laboratory activities:

Demonstration materials:

Notes for next time:

Chapter 5: Newton's Laws of Motion

Outline

5-1 Force and Mass
5-2 Newton's First Law of Motion
5-3 Newton's Second Law of Motion
5-4 Newton's Third Law of Motion
5-5 The Vector Nature of Forces: Forces in Two Dimensions
5-6 Weight
5-7 Normal Forces

Summary

Until now we have been studying the *effects* of motion (kinematics). Chapter 5 begins the study of the *causes* of motion (dynamics). An unbalanced force is one cause of motion. We will consider "normal sized" objects moving at "normal" speeds, keeping us in the realm of Newtonian physics. Newton's three laws are quite powerful and elegant and explain how an object moves when acted on by one or more forces. Chapter 5 introduces force as a *push* or *pull*. The vector nature of force is discussed and contrasted with the scalar nature of mass. Weight and the normal force are presented as examples of forces.

Major Concepts

By the end of the chapter, students should understand each of the following and be able to demonstrate their understanding in problem applications as well as in conceptual situations.

- Force
  - Vector nature of force
  - Weight
  - Normal force
- Mass
- Newton's laws
  - First law (law of inertia)
  - Second law ($\vec{\mathbf{F}} = m\vec{\mathbf{a}}$)
  - Third law (action-reaction force pairs)

Teaching Suggestions and Demonstrations

The challenge of this chapter is not learning Newton's laws; it is learning how to apply them to problems. Students often have difficulty with the vector nature of forces and with the difference between the concepts of weight and mass. This should not come as a surprise. Vectors are still new to these students. The rules for vector addition will have to be reinforced continually.

Section 5-1

Walk down any supermarket aisle and you will see that common usage confuses **mass** and **weight** (a **force**). Most food packages list the weight of the contents in pounds or ounces and the mass in kilograms or grams as if weight and mass were the same concept. (Why? The fundamental quantities in SI units are mass, time, and length; in British units they are force, time, and length. Besides, who wants to buy food by the slug, the unit for mass in the British system?) As the concepts are introduced, emphasize the physics definitions. Be sure to remind the students of the different definitions of mass and weight each time you work an example problem.

Mass is an intrinsic property of an object, the amount of matter contained in the object. Mass is a scalar. Units of mass are the kilogram (SI) or the slug (British units). The mass of an object stays constant (unless a piece falls off!) as the object moves from the Earth to another planet. Table 5-1 in the text gives a useful list of typical masses.

Force is a vector, with magnitude and direction. Units of force are the newton (SI) or the pound (British units). The forces that act on an object to cause motion are usually external to the object. (One exception is the case of explosions; another is the movement of the human body.)

Section 5-2

As stated in the text, **Newton's first law** is: "An object at rest remains at rest as long as no net force acts on it. An object moving with constant velocity continues to move with the same speed and in the same direction as long as no net force acts on it." This first law is sometimes called the Law of Inertia. It is important to note that in both cases above, whether the object is at rest or moving with constant velocity, the net force acting on the object is zero.

➲ **DEMO 5-1** The text briefly discusses the effects of friction on motion and describes the nearly frictionless environment of the air track. If you have access to an air track or an air table, set it up and let students suggest experiments with it to demonstrate the first law. (See Resource Information.)

If you don't have access to an air track, you can demonstrate Newton's first law with a series of small model cars. Start with one that has lots of friction in its axles. If you roll it across the floor, it will not go very far. Next find one that has very little friction in the axles. (Hot Wheels® work well and are inexpensive.) If you roll it across a smooth floor, it will go much farther than the first car. You can then talk about (theoretically) removing the friction entirely, and discuss what would happen.

You can also do a nice demonstration with an ice cube on one of the shiny black tabletops that are in many science classrooms. The ice surface melts slightly, so that the ice travels on a thin layer of water, a nearly frictionless situation.

This section also does a quick discussion of **inertial and non-inertial reference frames**. If you would like to add a little historical perspective, you can discuss Newton's idea of the absolute reference frame of the fixed stars. Since we now know that the stars are not really "fixed" in space, there are no absolute reference frames. We use the definition given in the text: If Newton's laws are true, then the reference frame is an inertial frame.

Encourage students to notice the many common examples of inertial and non-inertial reference frames, such as a car traveling at constant speed in a straight line, a car accelerating in a straight line, and a car traveling around a curve. What happens to a ball thrown up in the air in each case?

Section 5-3

Newton's second law can be stated as "unbalanced forces cause accelerations," or $\vec{\mathbf{a}} = \vec{\mathbf{F}}/m$. In its more familiar form, this is: $\Sigma\vec{\mathbf{F}} = m\vec{\mathbf{a}}$. Note that the sum of the forces is the sum of the *external* forces acting on the object, and that mass (NOT WEIGHT!) is the constant of proportionality between the net force and the acceleration. Point out that the acceleration is a *result* of the application of a force to a mass.

➲ DEMO 5-2 With a spring scale and several different masses, measure the amount of force required to pull each of the masses across a table top at approximately constant velocity. Does it take more force to pull a larger mass? (Friction is at work here!) (See Resource Information.)

Be sure to make the point that the equation given above is a vector equation that can be written as three independent scalar equations. Table 5-2 lists units of mass, acceleration, and force in the three main systems of units. Conceptual Checkpoint 5-1 gives a nice illustration of how Newton's laws can be applied to real-world situations.

Free-body diagrams are essential for solving problems using Newton's second law. Show your students lots of examples, such as those in Figure 5-5 and Example 5-1, and provide plenty of opportunities for practice. The problem-solving steps outlined in this section can be turned into another mantra:

- Draw a picture
- Isolate the object of interest
- Draw in the external forces acting on the object of interest
- Choose a coordinate system
- Resolve the forces into components
- Apply Newton's second law in each coordinate direction
- Solve the equations.

It is important to work many examples so that students can see how to apply these techniques in a variety of situations. Go through some examples that are in the text (Active Example 5-1 is a good one), but be sure to do some that are not worked in the text as well.

Section 5-4

Newton's third law states: "For every force that acts on an object, there is a reaction force acting on a different object that is equal in magnitude and opposite in direction." Students often have trouble with this concept. Remind them that the two forces act on *different* objects, so they appear in *different* free-body diagrams and therefore cannot cancel. Also, since the two forces act on different objects, they usually produce different accelerations. Example 5-3 is a good one to work in class.

Sections 5-5 through 5-7

Understanding the idea of **forces in two dimensions** is difficult for students who are not comfortable working with vectors. It is important to emphasize that motion in each coordinate direction is independent and that problems can be made easier by a careful choice of axis. You will have to remind students how to take the components of a vector and point out that the x and y components of a force can be associated with the sine *or* cosine of an angle, depending on which angle is chosen.

A discussion of **weight** presents the opportunity to demonstrate again the difference between weight (or force) and mass. Remind students that we sometimes use the terms mass and weight interchangeably in everyday language. This works because we are all confined to the surface of the Earth (or at least close to it!) and g is a constant in this region. If g changed from place to place, weight would also change, but mass would remain constant. (Remember, astronauts in space are "weightless" but not massless.)

There is a nice presentation of **apparent weight** in Figure 5-11 and the accompanying discussion. Note that if the object is resting on a horizontal scale, the apparent weight is the normal force.

➲ **DEMO 5-3** If your class or lab section is small enough, send them out in pairs with a bathroom scale to ride an elevator and measure **apparent weight**. If the students can travel between the same floors, you can use the measurements of the apparent and actual weights to calculate the acceleration of the elevator. If your class is too large, recommend this as an out-of-class assignment.

As you talk about apparent weight, you can lead right into a discussion of the **normal force**. Students usually catch on quickly to the concept of the normal force on a flat, horizontal surface; they have a little more trouble with an inclined or curved surface. Remind the students that the normal force is always perpendicular to the surface, though not always along the same axis as the weight. Go through Figure 5-15, which shows the components of the weight on an inclined plane, and do Example 5-9 to help clarify these points.

Resource Information

Transparencies

36. Table 5-1	Typical Masses in Kilograms (kg)
37. Table 5-3	Typical Forces in Newtons (N)
38. Figure 5-5	Constructing and using a free-body diagram
39. Figure 5-6	A book supported in a person's hand
40. Active Example 5-1	The Force Exerted by Foamcrete
41. Example 5-4	When Push Comes to Shove
42. Figure 5-9	Two astronauts pushing a satellite with forces that differ in magnitude and direction
43. Figure 5-10	Weight and mass
44. Figure 5-12	The normal force may equal the weight
45. Figure 5-13	The normal force may differ from the weight
46. Figure 5-15	Components of the weight on an inclined surface

Physlet® Illustrations

5-1 Force and Acceleration
5-2 Block Sliding on an Incline

Suggested Readings

Bao, L., Hogg, K., and Zollman, D., "Model Analysis of Fine Structures of Student Models: An Example With Newton's Third Law," *American Journal of Physics* (July 2002), pp. 766–778. A study of how the contextual features of problems affect student reasoning.

Bernhard, K. and Bernhard, J., "Mechanics in a Wheelchair," *The Physics Teacher* (December 1999), pp. 555-556. A description of a kinesthetic experience of Newton's laws.

Brand, H., "Action-Reaction at a Distance," *The Physics Teacher* (March 2002), pp. 136-137. Describes a procedure for demonstrating action-reaction in a case when the two bodies are not in contact.

Chandler, D., "Newton's Second Law for Systems with Variable Mass," *The Physics Teacher* (October 2000), p. 396. This example does involve calculus but is a good illustration of the variable mass problem.

Court, J.E., "Free-Body Diagrams Revisited - I," *The Physics Teacher* (October 1999), pp. 427-433. Free-body exercises with solutions in linear and circular motion.

Cross, R., "Standing, Walking, Running, and Jumping on a Force Plate," *American Journal of Physics* (April 1999), pp. 304-309. Details of an inexpensive force plate designed to measure ground reaction forces involved in human movement.

Gettrust, E., "An Extraordinary Demonstration of Newton's Third Law," *The Physics Teacher* (October 2001), pp. 392-393. Description of an apparatus using magnets and force probes to demonstrate that the action and reaction forces are equal in magnitude.

Haugland, O., "Physics Measurements for Sports," *The Physics Teacher* (September 2001), pp. 350-353. Contains instructions for making a simple force platform and for using ultrasonic motion detectors in helping athletes gain insight into the physics of their sports.

Linthorne, N., "Analysis of Standing Vertical Jumps Using a Force Platform," *American Journal of Physics* (November 2001), pp.1198-1204. Describes the use of a force platform to demonstrate the kinematics and dynamics of vertical jumping.

Mainardi, R., "Demonstration Experiments with Platform Scales," *The Physics Teacher* (November 2001), pp. 488-489. Describes three experiments, including one on Newton's third law.

Styer, D., "The Word 'Force'," *American Journal of Physics* (June 2001), pp. 631-632. A letter to the editor containing many examples to help students differentiate the physics meaning of "force" from its everyday meaning.

Wilczek, F., "Reference Frame: Mass without Mass I: Most of Matter," *Physics Today* (November 1999), pp. 11-13. Discusses where mass comes from.

Wilczek, F., "Reference Frame: Mass without Mass II: Most of Matter," *Physics Today* (January 2000), pp. 13-14. Deducing mass as a secondary property of matter.

Williams, K., "Inexpensive Demonstrator of Newton's First Law," *The Physics Teacher* (February 2000), p. 80. Uses a Downy® Ball fabric-softener dispenser!

Materials

Two air-track systems are available from Fisher Scientific: model number S52229, for use with a computer interface, or model number S52227A, for use as a stand-alone.

Spring scales are available from Ohaus in a variety of measurement ranges

A metal block set is also available from Fisher Scientific, model number S41245.

Notes and Ideas

Class time spent on material: Estimated:______ Actual:______________

Related laboratory activities:

Demonstration materials:

Notes for next time:

Chapter 6: Applications of Newton's Laws

Outline

6-1 Frictional Forces
6-2 Strings and Springs
6-3 Translational Equilibrium
6-4 Connected Objects
6-5 Circular Motion

Summary

Chapter 6 introduces more "real world" concepts to the study of forces. Frictional forces, both static and kinetic, are explained in terms of the microscopic interactions of surfaces. Strings transmit force along their length and can only pull on an object. (You can't push with a string!) Pulleys and springs are considered massless for now. Pulleys simply change the direction of the tension in a string. Springs obey Hooke's Law, $F_x = -kx$.

Major Concepts

By the end of the chapter, students should understand each of the following and be able to demonstrate their understanding in problem applications as well as in conceptual situations.

- Friction
    - Static friction
    - Kinetic friction
- String problems
    - Assumptions
    - Transmission of force
- Spring problems
    - Assumptions
    - Hooke's Law ($F_x = -kx$)
- Translational equilibrium ($\sum \vec{\mathbf{F}} = 0$)
- Motion of two connected objects
- Circular motion

Teaching Suggestions and Demonstrations

This chapter presents detailed applications of the concepts from Chapter 5. New types of forces and systems are introduced. As in Chapter 5, the students will need to see many examples. Be sure to work several that are not worked in the text.

Section 6-1

Most students do not have much trouble with the concept of **kinetic friction**. They experience it every day, so it seems familiar, and they have an intuitive understanding of the direction (opposing motion). **Static friction** is not as simple as kinetic friction. The inequality in the equation tends to confuse the students.

➲ **DEMO 6-1** Do the demonstration shown in Figure 6-3 with a block and a spring scale. Use several different blocks with different masses to show that the **force of static friction** can vary. Find the weight of each block and calculate the coefficient of static friction in each case. Then let the block move and determine the coefficient of kinetic friction. Which coefficient is larger?

Note: Accident reconstructionists will drag a tire across the roadway at the scene of an accident to determine the coefficient of kinetic friction between the tire and the road. This helps with determining how cars moved during the accident.

Table 6-1 lists typical coefficients of friction between common surfaces. Be sure to work problems in which the force of static friction is *not* at its maximum value to show students that the value of the force can vary.

Be prepared to discuss the "rule of thumb" that the frictional force is independent of the area of contact. This often comes up in the context of car tires, particularly race-car tires. You can point out that for a car traveling normally, the friction between the tires and the road is static friction. The tread expands and contracts as the tire surface contacts the road. This provides extra interactions between the tire and the road and so increases the coefficient of static friction. Conceptual Checkpoint 6-1 and the Real World Physics discussion on antilock braking systems are quite helpful.

Section 6-2

You may want to begin the discussion of **strings** by justifying the assumption that most strings can be considered massless.

➲ **DEMO 6-2** Hang a block from a **string**. Find the weight of the block and the weight of the string. Calculate the tension at the top of the string, where the string is supporting its own weight as well as the weight of the block, and at the bottom of the string, where just the weight of the block is supported. How different are these tensions? Are we justified in making the assumption that an ideal string is massless?

Remind the students that you can only pull (not push!) with a string.

Ideal **pulleys** are also considered massless and serve only to change the direction of the tension in the string. Example 6-4 is a good one to work in class, particularly if you have students interested in the pre-health professions.

Ideal **springs** are also considered massless. Springs obey Hooke's Law ($F_x = -kx$). You will need some examples of springs that can be compressed as well as stretched. These can be found in dart guns, retractable ballpoint pens, or pinball machines.

➲ **DEMO 6-3** Hang a spring vertically and note its hanging length. (See Resource Information.) Hang several different masses from it, one at a time, and note the change in the length for each mass. Calculate the **spring constant** of the spring.

Sections 6-3 and 6-4

Translational equilibrium combines ideas from Chapter 5 with new concepts from Chapter 6. Remind students that $\sum \vec{\mathbf{F}} = 0$ is a special case of $\sum \vec{\mathbf{F}} = m\vec{\mathbf{a}}$. Students will need to see many examples. Emphasize the importance of resolving vectors into their components, as in Example 6-5.

When two **connected objects** are in motion, how do they move? We begin by assuming that the string that connects the objects is inextensible. We can draw a free-body diagram for each object individually and then write separate equations of motion for each object. (Use the "mantra" from Chapter 5.) Since the objects are tied together, the magnitudes of the velocities and accelerations of the objects must be the same. It is possible for the directions of the velocity and acceleration to be different for the two objects, especially if pulleys are in the problem. (It is sometimes necessary to draw a free-body diagram for the pulley in order to solve a problem, as in Figure 6-18 for Problem 6-21. Here the block is attached directly to the pulley rather than to the rope.) Objects that are connected along a straight line can be treated as a single system, as shown in Figure 6-11.

Section 6-5

Centripetal motion is another concept that is not intuitive for students. The discussion at the beginning of Section 6-5 is excellent; it reminds the students that if an object is not moving in a straight line, it must have a force on it, and justifies the direction of the force. You will need to go through the derivation of Equation 6-14 carefully. There are several steps that are tricky for the students.

Emphasize that the **centripetal force** is not a mysterious force imposed on a situation from the outside. It must be provided by some force in the problem (*e.g.* gravity, the tension in a string, the normal force).

Any force that is serving as a centripetal force can be written in the form: $f_{cp} = ma_{cp} = m\frac{v^2}{r}$.

Example 6-8 involves a car going around a flat curve. It leads nicely into the Real World Physics discussion of skids and banked roadways.

➲ **DEMO 6-4** The traditional ball on a string swung in a horizontal circle works well for demonstrating **circular motion**. Variations include a (small) bucket of water or sand swung in a vertical circle (very dramatic!). See Conceptual Question 6-8.

Students like to see the mechanics of amusement park rides worked out. You can do the case of the loop-the-loop roller coaster (where the cars go upside down) and the spinning cylinder ride (see Conceptual Question 6-17). Have the students work together in pairs or groups of three to figure out why the riders don't fall out of the ride in either case!

The end of this chapter is a good time to schedule a problem session. If you have an extra class, lab time or a recitation time available, you can use that. Otherwise, schedule a problem session outside of class. The students need to see lots of examples, preferably before and after they have tried to work problems on their own.

Resource Information

Transparencies

47. Table 6-1	Typical Coefficients of Friction
48. Figure 6-2	Kinetic friction and the normal force
49. Example 6-2	Making a Big Splash
50. Figure 6-3	Static friction
51. Figure 6-6	Tension in a heavy rope
52. Figure 6-8	Spring Forces
53. Example 6-6	Connected Blocks
54. Example 6-7	Atwood's Machine
55. Figure 6-13	A particle moving in a circular path centered on the origin
56. Example 6-8	Rounding a Corner
57. Figure 6-15	A particle moving in a circular path with tangential acceleration

Physlet® Illustrations

6-1 Static and Kinetic Friction
6-2 Pulling a Block at an Angle
6-3 Spring Constant

Suggested Readings

Bouffard, K., "Physics Olympics: The Inertia Ball," *The Physics Teacher* (January 2001), pp. 46-47. A nice demonstration for centripetal force.

Charoenkul, N., Wheeler, D. and Dejasvanong, C., "The Wall of Death: Newtons, Nerves, and Nausea," *The Physics Teacher* (December 1999), pp. 533-535. A dramatic example of circular motion, centripetal force, and friction.

Court, J.E., "Free-Body Diagrams Revisited—I," *The Physics Teacher* (October 1999), pp. 427-433. Free body exercises with solutions in linear and circular motion.

Dalton, R., "Caught On Camera," *Nature* (15 August 2002), pp. 721–722. An interesting note on the use of high-speed cameras in biomechanics.

Kunzig, R., "The Physics of Walking: Falling Forward," *Discover* (July 2001), pp. 24-25. A fascinating general article on the physics involved in walking.

Larabee, D., "Car Collisions, Physics, and the State Highway Patrol," *The Physics Teacher* (September 2000), pp. 334-336. A real world application of the laws of motion and friction.

Larson, R., "Centrifugal Force and Friction," *The Physics Teacher* (October 1999), pp. 426-427. Uses a rotary motion probe to investigate circular motion.

Leonard, W., "Dragging a Box: The Representation of Constraints and the Constraint of Representations," *The Physics Teacher* (October 2001), pp. 412-414. A non-calculus derivation of the best angle to drag a box, along with a discussion of presentations of normal, friction, and tension forces.

Liphardt, J., Bibiana, O., Smith, S., Tinoco, I. and Bustamante, C., "Reversible Unfolding of a Single RNA Molecule by Mechanical Force," *Science* (27 April 2001), v. 292, pp. 733-737. A nice application of forces in biology.

Morrow, R., Grant, A. and Jackson, D., "A Strange Behavior of Friction," *The Physics Teacher* (October 1999), pp. 412-415. A more in-depth look at the subtleties of kinetic friction.

Newton, I. and Henry, R., "Circular Motion," *American Journal of Physics* (July 2000), pp. 637-639. Presentation of a simple derivation of the formula for the acceleration that occurs in uniform circular motion.

Reichert, J., "How Did Friction Get So 'Smart'?," *The Physics Teacher* (January 2001), pp. 29-31. An interesting discussion of the frictional force.

Ronhovde, P. and Sirochman, R., "Center of Mass Correction to an Error-Prone Undergraduate Centripetal Force Lab," *American Journal of Physics* (February 2003), pp. 185–188. Describes a simpler correction to a centripetal force laboratory experiment.

Sawicki, M., "Static Elongation of a Suspended Slinky™," *The Physics Teacher* (May 2002), pp. 276-278. A precalculus treatment of the statics of a loosely wound spring, using both mathemathical summation techniques and physical considerations.

Van den Berg, W., "The Best Angle for Dragging a Box," *The Physics Teacher* (November 2000), pp. 506-508. A great exercise in calculation of vector components and friction. Requires a graphing calculator or computer.

Materials

A Hooke's law spring set is available from Sargent-Welch, model number CP33855-00.

Notes and Ideas

Class time spent on material: Estimated:______ Actual:_____________

Related laboratory activities:

Demonstration materials:

Notes for next time:

Chapter 7: Work and Kinetic Energy

Outline

Summary

Chapter 7 introduces students to the important concepts of work and energy. Kinetic energy, or energy of motion, is discussed in detail. Power, the time rate of change of doing work, is also defined. After the foundation laid in Chapter 7, students are ready to discuss potential energy, which is another form of mechanical energy, as well as the law of conservation of energy in Chapter 8.

Major Concepts

By the end of the chapter, students should understand each of the following and be able to demonstrate their understanding in problem applications as well as in conceptual situations.

- Work
    - Force in the direction of displacement
    - Force at an angle to displacement
    - Positive, negative and zero work
    - Constant force and variable force
- Kinetic energy
- Work-energy theorem
- Power

Teaching Suggestions and Demonstrations

Throughout this chapter, students encounter terms that have different meanings in English than in physics. Emphasize to students that the words "work," "energy," and "power" have specific definitions in physics that are somewhat related to but definitely not exactly the same as their meanings in everyday conversations. Point out that when they say they've "worked" really hard on their homework, they are speaking English and not physics!

Sections 7-1 and 7-2

Students may at first be confused to find that they do no **work** in trying to push a stalled car that won't move or in holding a heavy box up in the air. Direct them to the discussion in Section 7-1 to reassure them that on the microscopic level, the cells in their muscles are contracting and expanding and therefore doing work. However, no net work is done on the car or the box. The weight lifter shown in the photograph is a good example. How much work does she do *holding* the 150 kg over her head once she has raised it there? None.

This section also provides a good opportunity to point out that numbers don't always tell the whole story when it comes to **work** and **force**. The man pulling the suitcase in Figure 7-2 could do the same amount of work with less force if he pulled the suitcase horizontally. Why doesn't he? Well, for one, walking through an airport on your knees is uncomfortable! You can also discuss what the vertical component of his force is accomplishing. It reduces the normal force on the suitcase, which doesn't contribute to the work done but which does reduce friction if any is present. You can extend Conceptual Checkpoint 7-1 for another example. If no friction is present, the same amount of work is done in lifting the box straight up or in sliding it up the ramp. If friction is present, more work is needed to slide it up the ramp. Even though real-life ramps have friction, most of us would choose to do more work and use the ramp if we are moving heavy boxes. Why? Because we can apply less force. Although power has not yet been introduced, you can also point out that less power is needed to push the box up the ramp. Then return to this example when Section 7-4 is discussed.

The SI unit of work and energy, the **joule**, will most likely be unfamiliar to students. Table 7-1 and Exercises 7-3 and 7-4 will help them get a feel for just how much work (or energy) one joule is.

➲ **DEMO 7-1** To demonstrate the relationships among **work, force,** and **displacement**, attach a block to a spring scale and use the scale to drag the block across a table. The work performed by the scale on the block is equal to the component of force parallel to the tabletop times the distance the block moves along the table. (In order to actually do the calculations, you will need to pull the block so that the force stays constant.) Students can try this exercise in small groups or you can have a few demonstrate to the whole class and then involve the others in the calculations. Try pulling the block with a horizontal force and then with a force applied at an angle.

When introducing the **work-energy theorem**, emphasize that it is the **total work** done on an object that is equal to its change in kinetic energy. Figure 7-4 explains how work can be positive, negative or zero. Give some examples of **negative work**, such as the work done in catching a baseball or braking a car. Conceptual Questions 7-7, 7-8, 7-9, and 7-15 at the end of the chapter provide a check of student understanding of negative work.

➲ **DEMO 7-2** The above demonstration can be extended to illustrate the **work-energy theorem**. Try pulling the block with just the constant force necessary to keep it moving at a constant velocity. Then apply a larger constant force and note that it speeds up. In the first case, the change in KE is zero, so the total work done is zero. Friction did an amount of **negative work** on the block just equal to the positive work done by the spring scale. (If you have time, you can even use the exercise to calculate the coefficient of friction between the block and the table.) In the second case, the change in KE is positive as is the net work done on the block.

Students may at this point be tempted to think of work as a vector. It isn't. Work is a **scalar** that can have positive, negative or zero values depending on the angle between the force and the displacement.

➲ **DEMO 7-3** If you feel particularly adventuresome, an egg-catching demonstration is a great way to investigate both the **work-energy theorem** and the relationships among **work**, **force**, and **distance**. Toss an egg to a student and have the class discuss why she moves her hand back in catching it. An important point for them to understand is that the catcher has no control over the amount of work she has to do. (The work-energy theorem says that her work equals the change in KE; the egg has a certain initial speed and the catcher wants to bring it to rest, so the change in KE is fixed.) However, she can control force and displacement. The same work will be accomplished by applying a small force over a large distance (i.e. pulling her hand back,) or a large force over a small distance (splat!).

The same relationships can be demonstrated with less mess by catching baseballs. It hurts a lot less to move your hand back with the ball because your force on the ball is less and therefore, according to Newton's third law, the ball's force on your hand is less as well.

Section 7-3

In the case of a **variable force**, work done is the **area under the curve** on a force-position graph. Students may be confused by the fact that an area has units of joules instead of square meters. Refer to the footnote in this section and to Figure 7-6. Figure 2-14 from Chapter 2 can be revisited to remind them of another instance when the area under a curve has important physical significance; the area under the curve on a velocity-time graph is displacement. If any students in your class have had introductory calculus, point out that Figure 7-8 is showing an integral.

Students are very likely to assume that it will take twice as much work to stretch a **spring** twice as far. Example 7-7 is a good problem to do in order to address this misconception and to continue the discussion of force and work for the specific case of a spring. Conceptual Question 7-11 at the end of the chapter is a comparable question regarding the work done in accelerating a car.

Section 7-4

Power is work (or energy) per unit time, or the rate of doing work. Introduce the SI unit of power, the **watt**, and contrast it with a kilowatt-hour, the unit that typically appears on a "power" bill. Point out that people actually pay the power company for energy consumed, not power, since a kilowatt-hour (or power times time) is a unit of energy. Do an example converting a typical monthly energy usage in kilowatt-hours to joules so that students will begin to get a feel for the relationship between these two energy units. As more and more new units are introduced in the course, it is helpful to occasionally do a "units check" of an equation. Two equations are given for power in this section, work per time and force times speed. Have students show that the units from these two expressions are the same.

➲ **DEMO 7-4** Problem 7-36 at the end of the chapter involves calculating the **power output** of a man running up the steps of the Empire State Building. Your students can run up bleachers or a flight of steps and calculate their own power outputs.

➲ **DEMO 7-5** If you go refrigerator shopping, you will probably find **energy efficiency ratings** for various models prominently displayed. Bring in prices and ratings of a few different refrigerators, as well as a local power bill so you will know the cost of electricity per kilowatt-hour. Students can estimate the monthly cost of running a refrigerator, and if they find that the more efficient models are more expensive to buy than the less efficient models, they can also estimate the payback time for investing in the more efficient model.

Active Example 7-2 discusses the maximum speed a car can go up a hill given the horsepower delivered by the engine. Extend this example by discussing why trucks are so slow going up hills. Since the mass of the truck is much greater, the force needed to keep it moving up a hill is also much greater.

Resource Information

Transparencies

58. Table 7-1 Typical Values of Work

59. Figure 7-2	Work: force at an angle to direction of motion
60. Conceptual Checkpoint 7-1	
61. Figure 7-4	Positive, negative, and zero work
62. Example 7-3	A Coasting Car I
63. Figure 7-7	Work done by a non-constant force
64. Figure 7-8	Work done by a continuously varying force
65. Figure 7-9	Stretching a spring
66. Figure 7-11	Work done in stretching a spring: Average force
67. Figure 7-12	The work done by a spring can be positive or negative
68. Figure 7-13	Driving up a hill

Physlet® Illustrations

7-1 Work and Angles
7-2 Work on an Incline

Suggested Readings

Hilborn, R., "Let's Ban *Work* from Physics," *The Physics Teacher* (October 2000), p. 447. A commentary on problematic terminology.

Ingham, W., "A Consistent Sign Convention for Work," *The Physics Teacher* (March 2000), p. 160. Discussion of positive and negative sign conventions for work.

Malone, J. and Holzwarth, D., "A Real Look at Speeds and Stopping Distances," *The Physics Teacher* (February 1998), pp. 95-96. A real world demonstration of the relationship of speed and stopping distance.

Mendelson, K., "Physical and Colloquial Meanings of the Term 'Work'," *American Journal of Physics* (March 2003), pp. 279–281. An interesting article on the history of the term 'work'.

Piatek, S. and Gautreau, R., "Constant Acceleration and Kinetic Friction," *The Physics Teacher* (May 1998), p. 316. A very simple experiment connecting work, acceleration, and kinetic friction.

Velázquez-Avilés, A., "Using the Work–Energy Theorem with *Car and Driver* Website Data," *The Physics Teacher* (April 2002), pp. 235-237. A great exercise to illustrate the work-energy theorem using real data.

Notes and Ideas

*Class time spent on material: Estimated:*______ *Actual:*_____________

Related laboratory activities:

Demonstration materials:

Notes for next time:

Chapter 8: Potential Energy and Conservative Forces

Outline

8-1 Conservative and Nonconservative Forces
8-2 Potential Energy and the Work Done by Conservative Forces
8-3 Conservation of Mechanical Energy
8-4 Work Done by Nonconservative Forces
8-5 Potential Energy Curves and Equipotentials

Summary

Chapter 8 continues the discussion of energy by introducing potential energy. The important distinction between conservative and nonconservative forces is made. Conservation of mechanical energy is discussed in detail. Solving problems using energy considerations often turns out to be easier than solving them with kinematic equations.

Major Concepts

By the end of the chapter, students should understand each of the following and be able to demonstrate their understanding in problem applications as well as in conceptual situations.

- Conservative and nonconservative forces
  - Work and stored energy
  - Path dependence or independence of work
- Potential energy
  - Gravitational
  - Spring
  - Potential energy curves and equipotentials
- Conservation of mechanical energy
- Work done by nonconservative forces; changing mechanical energy
- Law of conservation of energy

Teaching Suggestions and Demonstrations

This chapter is your first chance to show the power of conservation laws in physics. Plan to work several problems using methods from earlier chapters, then work them again using energy conservation methods. The (usually) dramatic difference between the complexity of the first solution and the simplicity of the second is enough to convince most students to learn the new method.

Sections 8-1 and 8-2

The distinction between **conservative and nonconservative forces** is made clearly in the first section. Table 8-1 categorizes the forces the students have been introduced to so far as either conservative or nonconservative. Two definitions of conservative force are given in Section 8-1; show students that these definitions are consistent with the fact that energy can be stored and recovered by conservative forces. For instance, a pile driver is simply a large mass raised in a gravitational field. Energy put into the system raises it in a gravitational field; energy is released in the form of work done on the pile when the mass falls.

The two conservative forces addressed directly are the **spring force** and **gravity**. In both cases, **potential energy** of the system in a certain state is equal to the work done in order to put it in that state. Students are already familiar with the formulas for work needed to compress or stretch a spring and work needed to raise an object in a uniform gravitational field. Point out that the equations for potential energy are the same.

Sections 8-3 and 8-4

Conservation of mechanical energy is the first conservation law encountered in physics, so it is important to ensure that students understand what "conservation" means. If a quantity is conserved, then the total value of that quantity remains constant. Conservation of energy in physics refers to the fact that energy isn't created or destroyed, although it does change in form. Kinetic energy is converted to potential energy when a book is tossed up into the air. Kinetic energy is converted into heat when a book slides across a table and comes to rest due to friction. Mechanical energy (potential plus kinetic) is conserved only in the first case (when only conservative forces are acting), although total energy is conserved in both. The difference is that the heat energy cannot be easily converted back to organized, macroscopic motion; there is no way to gather the energy back out of the particles and organize it to spontaneously make the book move again. (Note that "energy conservation" in environmental science means something entirely different!)

➲ **DEMO 8-1** A pendulum and a spring are both simple demonstrations of **conservation of mechanical energy**. Set a mass on a spring oscillating and a pendulum swinging and point out when the energy is all kinetic or all potential. By measuring the vertical displacement from equilibrium of either the mass on the end of the pendulum or the mass on the end of the spring, students can calculate the velocity the mass has as it passes through equilibrium. In both cases, the mechanical energy is not perfectly conserved, as is made clear by the fact that the mass eventually stops swinging or stops bobbing up and down. Ask students what nonconservative forces (friction and air resistance) may be acting on the system.

Example 8-5 uses kinematics equations from earlier chapters to solve a motion problem and then demonstrates that the total mechanical energy at different points in the motion is the same. This example gives students direct evidence for the **conservation of mechanical energy,** and they are subsequently encouraged to use energy considerations to solve problems.

The examples in Section 8-4 illustrate how energy considerations can be useful even if mechanical energy is not conserved, as is the case when **nonconservative forces** are acting. Compare Example 8-10 to Example 6-6, which consists of a similar configuration of connected blocks. Students can also solve the problem in Example 8-10 using Newton's laws; help them discover that using the change in mechanical energy is easier.

➲ **DEMO 8-2** **Roller coasters** provide excellent illustrations of kinetic and potential energy and the conservation of mechanical energy. Demonstration setups complete with photogates for determining velocities are available for purchase. (See Resource Information.) Alternatively, you could use a toy car, train, or marble and an appropriate track. Comparison of initial height and final height will allow students to determine the total mechanical energy "lost" due to nonconservative forces and the coefficient of friction between the track and the moving object. In addition, there are numerous web sites devoted to roller coaster design and other applications of physics in amusement parks. (See Resource Information.)

Section 8-5

A simple roller-coaster track is essentially a **potential energy curve**, so the demonstration above leads nicely into the last section of the chapter. Draw a hypothetical two-dimensional roller-coaster track on the board and point out, as in Figure 8-11, that since height is proportional to potential energy, the curve of the track is the same as the curve on a potential-energy versus horizontal-position graph. The potential energy curve for a mass on a spring (Figure 8-12) is a little harder to visualize. Return to the comparison between the mass on the spring and the pendulum to help students understand this graph.

Resource Information

Transparencies

69. Figure 8-3	Work done by gravity on a closed path
70. Figure 8-4	Work done by friction on a closed path
71. Example 8-4	Compressed Energy
72. Figure 8-8	Solving a kinematics problem using conservation of energy
73. Figure 8-9	Speed is independent of path
74. Conceptual Checkpoint 8-1	
75. Example 8-8	Spring Time
76. Example 8-10	Landing with a Thud
77. Figure 8-10	A ball rolling on a frictionless track
Figure 8-11	Gravitational potential energy versus position for the track shown in Figure 8-10
78. Figure 8-12	A mass on a spring

Physlet® Illustrations

8-1 Roller Coaster
8-2 Work Done by Friction

Suggested Readings

Arons, A., "Development of Energy Concepts in Introductory Physics Courses," *American Journal of Physics*, (December 1999), pp. 1063-1067. Discussion of the use of work in conservation-of-energy calculations.

Ehrlich, R., "Using a Retractable Ball Point Pen to Test the Law of Conservation of Energy," *American Journal of Physics* (February 1996), p. 176. A clever demonstration/experiment that uses the elastic potential energy of the compressed spring from a pen to launch a projectile.

Harrison, E., "Mining Energy in an Expanding Universe," *The Astrophysical Journal* (10 June 1995) v. 446, pp. 63-66. An intriguing look at whether or not energy is conserved on a cosmic scale in an expanding universe.

Jolly, P., Zollman, D., Rebello, N. and Dimitrova, A., "Visualizing Motion in Potential Wells," *American Journal of Physics*, (January 1998), pp. 57-63. Description of a unit designed to aid learning of potential-energy diagrams that incorporates a sequence of computer-interfaced experiments using dynamics or air-track systems.

Keeports, D., "How Does the Potential Energy of a Rising Helium Balloon Change?" *The Physics Teacher* (March 2002), pp. 164-165. Considers gravity and the buoyant force, two conservative forces acting on a rising helium balloon.

Tanner, R., "Roller-Coaster Design Project," *The Physics Teacher* (March 1997), pp. 148-149. Description of a project requiring conservation of energy and other physics concepts, including a sample handout of the assignment.

Taylor, J., Carpenter, A. and Bunton, P., "Conservation of Energy with a Rubber Ramp," *The Physics Teacher* (March 1997), pp. 146-147. Instructions for construction of a conservation of energy lab and demonstration apparatus.

http://www.learner.org/exhibits/parkphysics/ is a web site devoted to amusement park physics and includes roller coasters and pendulum rides. It is maintained by The Annenberg/CPB Project. A Web search for "roller-coaster design" will lead you to numerous other related sites.

Materials

AAPT's Official Guide: Amusement Park Physics, available from www.aapt.org/catalog or by calling 301-209-3333.

One roller-coaster apparatus is available from Frey Scientific, model number S618337.

Drilled balls for use as pendulum bobs are also available from Frey Scientific.

Notes and Ideas

Class time spent on material: Estimated:______ *Actual:______________*

Related laboratory activities:

Demonstration materials:

Notes for next time:

Chapter 9: Linear Momentum and Collisions

Outline

9-1 Linear Momentum
9-2 Momentum and Newton's Second Law
9-3 Impulse
9-4 Conservation of Linear Momentum
9-5 Inelastic Collisions
9-6 Elastic Collisions
9-7 Center of Mass
*9-8 Systems with Changing Mass: Rocket Propulsion

Summary

Chapter 9 introduces the concepts of linear momentum and impulse. Momentum is a vector quantity. The conservation of linear momentum is extremely important in the treatment of collisions, both elastic and inelastic. The center of mass is defined. The case of variable mass is covered as an optional topic.

Major Concepts

By the end of the chapter, students should understand each of the following and be able to demonstrate their understanding in problem applications as well as in conceptual situations.

- Linear momentum
  - $\vec{\mathbf{p}} = m\vec{\mathbf{v}}$
  - General form of Newton's second law ($\sum \vec{\mathbf{F}} = \Delta\vec{\mathbf{p}} / \Delta t$)
- Impulse ($\vec{\mathbf{I}} = \vec{\mathbf{F}}_{\mathbf{av}} \Delta t = \Delta\vec{\mathbf{p}}$)
- Conservation of momentum ($\sum \vec{\mathbf{F}} = \Delta\vec{\mathbf{p}} / \Delta t = 0$)
  - Internal and external forces
  - Recoil
- Collisions
  - Inelastic
  - Elastic
- Center of mass

Teaching Suggestions and Demonstrations

Most students do not have an intrinsic feel for the concept of linear momentum. You will need to work many examples comparing the momenta of everyday objects. The conservation of momentum is the second major conservation law that students encounter. Be sure to discuss the usefulness of collision theory in many areas of physics.

Sections 9-1 through 9-3

Linear momentum is defined as $\vec{\mathbf{p}} = m\vec{\mathbf{v}}$, with units kg m/s. The term "linear momentum" is often shortened to "momentum." Point out that the combination $m\vec{\mathbf{v}}$ appears in Newton's second law, in its original form ($\sum \vec{\mathbf{F}} = \Delta\vec{\mathbf{p}} / \Delta t$), and that it comes up often enough to earn a special name. Work several examples to illustrate the trade-off between mass and velocity in the calculation of momentum. It is very important to remind students that momentum is a **vector quantity**. Point out that the **net momentum of a system of objects** is just the vector sum of the momenta of all the objects in the system.

Use Figure 9-1 and go through the calculation in the text to illustrate the vector nature of the **change in momentum**. **Impulse** is defined as $\vec{\mathbf{I}} = \vec{\mathbf{F}}_{av}\Delta t = \Delta\vec{\mathbf{p}}$. Students also find this concept non-intuitive. Why do we need a separate quantity for the change in momentum? Why does it have the same units as momentum? Again, the justification can come from Newton's second law. Be sure to go over the concept of average force (see Figures 9-2 and 9-3) and work several examples. Talk about the pictures of the softball and the pole vaulter in Section 9-3.

It is also interesting to talk about impulse and momentum in relation to **traffic laws**. Why are pedestrians required to walk facing car traffic, but bicyclists required to ride in the same direction as car traffic? (It is in the best interest of a pedestrian to see traffic as it approaches; the momentum of a pedestrian is negligible compared to that of a moving car. It is in the best interest of a bicyclist to move in the same direction as the cars; if the bicyclist is hit, the impulse delivered by the car is less if the car and bicycle are moving in the same direction than if they meet head-on.)

You can also have a good class discussion about impulse and the **safety features for cars** (or bicycle helmets). Note that LARGE FORCE x short time = smaller force x LONGER TIME. Crumple zones in cars and the padding in helmets are all there to increase the time over which a collision will take place, thus reducing the force necessary for the same impulse. Refer back to the egg-catching Demo suggested for Chapter 7 as well. In that case we discussed the work done to stop an egg. If we increase the distance over which the egg is stopped, we can decrease the force required for stopping and yet do the same amount of work. In the current discussion, impulse stays constant, and increasing the stopping time decreases the force required.

Section 9-4

Conservation of momentum ranks as one of the most powerful ideas in physics. The full statement of the law of conservation of momentum is: If the net force acting on an object is zero, its momentum is conserved. It is important at this point to distinguish between **external** and **internal forces** acting on an object or a system of objects. The internal forces acting on a system will always sum to zero (Newton's third law). In order for momentum to be conserved, the external forces must also sum to zero.

Go through Example 9-3. Point out that in this case the *system* is made up of both canoes. As one canoe moves to the right, the other canoe **recoils** to the left. Recoil is also evident when throwing a ball on ice or when turning on a hose. This section contains a nice discussion of the recoil of the human body when the heart pumps blood.

Sections 9-5 and 9-6

Collisions are an important tool for finding out about the atomic and subatomic world. They also have many macroscopic applications. Collisions are divided into two general types: **inelastic collisions**, which conserve only momentum, and **elastic collisions**, which conserve momentum and kinetic energy. The special case of a one-dimensional, completely inelastic collision, in which the objects stick together after the collision, is the simplest and should be treated first. If the masses of the two colliding objects and their initial velocities are known, the only unknown is the final velocity of the combined objects.

Analyses of simple (rear-end) traffic collisions are interesting to the students and can give object lessons on the importance of wearing a seat belt.

If you choose to cover **two-dimensional collisions**, we recommend that you start with the simple case of a completely inelastic collision between objects with perpendicular initial velocities.

Analysis of one-dimensional elastic collisions involves the simultaneous solution of two equations, one of which is second order in velocity. Take the time to work out the equations for the final velocities (Equation 9-12) and go over the special cases illustrated in Figure 9-7. These can help the students develop intuition about collisions.

➲ **DEMO 9-1** An air track (or an air table) is an excellent demonstration device for collisions. Set up a couple of photogates for timing purposes and use carts with different masses to explore a variety of cases. Use metal spring bumpers on the carts to simulate elastic collisions and Velcro to simulate inelastic collisions. Demonstrate a variety of elastic and inelastic collisions.

➲ **DEMO 9-2** Collect three different sizes of superballs. Stack them on top of each other, with the smallest on top, and drop them together. (This takes some practice, but the result is spectacular!) Careful analysis of the collisions, assuming that the approximate ratio of the masses is 1:2:4, determines that the smallest ball bounces to nine times the original height. (The largest ball impacts with the floor and hits the middle ball, which then collides with the smallest ball. Since the balls fall about the same distance, they all have the same speed just before the initial impact.) (Caution – DO NOT DO THIS in a room with a low ceiling. In that case, use only two of the balls.)

➲ **DEMO 9-3** A Newtonian collision ball apparatus is a fairly easy and inexpensive way to demonstrate pendulum motion as well as conservation of energy and conservation of linear momentum. (See Resource Information.)

Section 9-7

At the beginning of this section, the **center of mass** of a system of masses is defined as "the point where the system can be balanced in a uniform gravitational field."

➲ **DEMO 9-4** Cut a random shape out of cardboard. Punch two small holes in it near the edge at positions about 120° from each other. Suspend the cardboard from one hole, hang a plumb bob from the hole, and mark a vertical line. Repeat the process for the second hole. The point where the two lines intersect is the center of mass of the shape. The shape can be balanced horizontally at that point.

For two or more objects, the position of the center of mass can be calculated as the weighted average of the positions of the objects:

$$x_{CM} = \frac{\sum m_i x_i}{M} \qquad y_{CM} = \frac{\sum m_i y_i}{M}$$

You can explain briefly (for any students who have had calculus) that these sums become integrals for the case of continuous objects. Note that the velocity and acceleration equations follow this same form, as does Newton's second law.

➲ **DEMO 9-5** Throw several extended objects (carefully) across the front of the room. Many common objects (erasers, pens, etc.) work well. A baton (or a meter stick) is particularly easy to see. Be sure to give each object some rotation about the center of mass as you throw it. Have the students watch the center of mass. They should be able to see it follow a parabolic path in each case, even as the rest of the object rotates around it.

*Section 9-8

This section is optional. We recommend discussing it qualitatively. The last paragraph in the section contains a clear explanation of how rockets move in outer space.

Resource Information

Transparencies

79. Figure 9-2	The average force during a collision
80. Figure 9-3	Hitting a baseball
81. Figure 9-5	Railroad cars collide and stick together
82. Example 9-5	Ballistic Pendulum
83. Example 9-6	Bad Intersection: Analyzing a Traffic Accident
84. Figure 9-6	An elastic collision between two air carts
85. Figure 9-7	Elastic collisions between air carts of various masses
86. Figure 9-8	Two curling stones undergo an elastic collision
87. Figure 9-10	The center of mass of two objects
88. Example 9-8	Center of Mass of the Arm
89. Figure 9-13	Center of mass of an exploding rocket
90. Figure 9-14	Weight and acceleration of the center of mass

Physlet® Illustrations

9-1 Elastic and Inelastic Collisions
9-2 Two-Dimensional Collisions
9-3 Where is the Center of Mass?
9-4 Rocket

Suggested Readings

Cross, R., "The Bounce of a Ball," *American Journal of Physics* (March 1999), pp. 222-227. A study of the dynamics of bouncing for several common ball types, including a discussion of energy loss during each collision and differing coefficients of restitution.

DeYoung, P. and Mulder, B., "Studying Collisions in the General Physics Laboratory With Quadrature Light Emitting Diode Sensors," *American Journal of Physics* (December 2002), pp. 1226–1230. Describes a technique to measure position as a function of time in the general physics lab.

Edge, R., "The Physics of Crème Brulée," *The Physics Teacher* (October 2000), p. 441. An interesting application of an impulsive force.

Ha, M., Kim, Y. and Lee, S., "Development of an Apparatus for Two-Dimensional Collision Experiments Using a Cycloidal Slide," *American Journal of Physics* (November 2001), pp.1187-1190. Describes an apparatus for performing high precision experiments of conservation of linear momentum in two-dimensional collisions.

Hu, H., "More on One-Dimensional Collisions," *The Physics Teacher* (February 2002), p. 72. A rearrangement of the equations for one-dimensional collisions that results in an alternate definition of the coefficient of restitution.

Loveland, K., "Simple Equations for Linear Partially Elastic Collisions," *The Physics Teacher* (September 2000), pp. 380-381. A nice treatment of inelastic collisions that are not completely inelastic.

Millet, L., "The One-Dimensional Elastic Collision Equation: $v_f = 2v_c - v_i$," *The Physics Teacher* (March 1998), p. 186. An interesting proposal for a new treatment of one-dimensional collisions.

Nathan, A., "Dynamics of the Baseball-Bat Collision," *American Journal of Physics* (November 2000) pp. 979-990. A complete description of the collision process including bending vibrations of the bat.

Penner, A. R., "The Physics of Golf: The Optimum Loft of a Driver," *American Journal of Physics* (May 2001), pp. 563-568. Detailed consideration of the motion of a golf ball, from the impact with the club to the run after landing.

Rist, C., "Breaking Boards," *Discover* (May 2000), pp. 28-31. A very accessible article on the application of physics principles to karate.

Turner, W. and Ellis, G., "The Energetics of a Bouncing Ball," *The Physics Teacher* (November 1999), pp. 496-498. An experiment involving kinetic and potential energy of a ball during collisions with the ground.

Van den Berg, E., Nuñez, J., Guirit, A. and Van Huis, C., "Cotton Buds, Momentum, and Impulse," *The Physics Teacher* (January 2000), pp. 52-53. An interesting experiment demonstrating impulse and momentum.

Materials

See air-track listing in Chapter 5.

A Newtonian collision ball apparatus is available from Fisher Scientific, model number S40976.

Notes and Ideas

Class time spent on material: Estimated:______ *Actual:______________*

Related laboratory activities:

Demonstration materials:

Notes for next time:

Chapter 10: Rotational Kinematics and Energy

Outline

Summary

Chapter 10 begins the study of rotational motion. The variables of rotational motion are defined, used in kinematic equations, and connected to their linear counterparts. A description of rolling is presented. The moment of inertia and rotational kinetic energy are defined, and energy considerations are discussed.

Major Concepts

By the end of the chapter, students should understand each of the following and be able to demonstrate their understanding in problem applications as well as in conceptual situations.

- Angular variables
    - Angular position θ
    - Angular velocity ω
    - Angular acceleration α
- Equations for rotational kinematics
- Connections with linear variables
- Rolling
- Rotational kinetic energy
- Moment of inertia
- Conservation of mechanical energy

Teaching Suggestions and Demonstrations

The mathematical description of angular motion is sometimes difficult for students to understand. Use lots of demonstrations, pictures, and diagrams. Point out the parallels and connections between linear and rotational motion.

Sections 10-1 through 10-3

➲ **DEMO 10-1** It is a good idea to have a bicycle wheel handy for this entire chapter. (See Resource Information.) Since the demise of the turntable, students often have trouble visualizing angular motion. You can demonstrate the angular variables easily and dramatically with a bicycle wheel. Use chalk to mark on the rim, then move the wheel through an angle θ. Tie a piece of red ribbon or yarn to a spot on the rim, and another to a spot on one of the spokes to demonstrate the difference in tangential velocity at different distances from the center of rotation.

As you discuss the **angular position**, you will need to go over the concepts of **radians**, the **sign conventions for angular position**, **conversions (radians**, **revolutions**, and **degrees)**, and **arc length**. Most students have covered these topics before, but most will not be familiar enough with them to use them easily.

There are many definitions and small equations in this chapter. Have students begin an equation card to help keep them organized. (A 3" x 5" card is fine.) Include the definitions for ω, α, the period T, tangential speed v, centripetal acceleration, and tangential acceleration. Remind the students of the value of dimensional analysis to help deduce the form of these equations.

Refer to the charts of the **correspondences between linear variables and angular variables** and between the linear kinematic equations and the angular kinematic equations in Section 10-2. These will aid in making the connections between the variables and in recognizing that these "new" equations aren't really "new"; they are just restatements of the familiar kinematic equations. It is important to point out that the angular quantities θ, ω, and α are the same for all parts of a rotating rigid object. The linear quantities x, v, and a will be different for different parts of the object.

Take time to work several kinematic examples. The students will need to see how various quantities fit into the solutions of problems, especially when the angular variables are given in different units (revolutions, radians, degrees).

Be sure to discuss the two **components (centripetal** and **tangential) of acceleration**. Students who have become used to one-dimensional acceleration will need extra help with this concept.

Section 10-4

Rolling is a combination of rotational motion and translational motion.

➲ **DEMO 10-2** Use the bicycle wheel to demonstrate **rolling**. Mark a point on the rim of the wheel, and roll the wheel along a tabletop. Ask a series of questions. What kind of motion is the center of mass of the wheel experiencing? What about the point on the rim? How does the point on the rim move in the vertical direction? In the horizontal direction? When is it moving fastest? Slowest? This will help the students develop an intuitive feel for rolling and make the mathematical treatment easier to understand.

Figure 10-11 gives a good illustration of the **horizontal velocity vectors** for different points on the wheel. Have the class look at the photograph of the rolling wheel and discuss why the blurring of the image occurs.

Section 10-5

Justifying the **moment of inertia** without calculus is difficult. The discussion at the beginning of this section is quite good. Talk about the idea that the moment of inertia gives information about the distribution of mass about an axis of rotation. The farther the mass is from the axis, the larger the moment of inertia. The larger the moment of inertia, the greater the resistance to change in rotational motion. For instance, tightrope walkers hold long poles to increase their moments of inertia. Gymnasts on balance beams hold out their arms for the same reason. Try balancing a tennis racket (or hammer) on your fingertip. It is easier to balance with the heavy end up, away from your fingertip.

You can use the correspondence of the angular quantity I to the translational quantity m to deduce the form of the **rotational kinetic energy**. Go over the moments of inertia listed in Table 10-1. It helps if you have examples of these shapes that you can use for models. Go through Conceptual Checkpoint 10-2 and discuss how the result is consistent with the information given in Table 10-1.

Section 10-6

The inclusion of the rotational kinetic energy in the **total mechanical energy** of a rolling object seems reasonable but is often not obvious to students. Use Conceptual Checkpoints 10-3, 10-4, and 10-5 to help illustrate this concept.

➲ **DEMO 10-3** Set up the apparatus for Conceptual Checkpoint 10-4 and do the experiment. You can make this into a real competition that the students will enjoy and remember. Use the disk and hoop, but also use spheres and small cars. Have the students calculate the moment of inertia for a toy car (remember, only the wheels rotate!) and predict which car or other object will get to the bottom of the incline the fastest.

Resource Information

Transparencies

91.	Figure 10-2	Arc length
92.	Figure 10-5	Angular acceleration
93.	Figure 10-6	A pulley with constant angular acceleration
94.	Figure 10-7	Angular and linear speed
95.	Figure 10-8	Centripetal and tangential acceleration
96.	Figure 10-9	Rolling without slipping
97.	Figure 10-10	Rotational and translational motion of a wheel
98.	Figure 10-11	Velocities in rolling motion
99.	Figure 10-14	A dumbbell-shaped object rotating about its center
100.	Table 10-1	Moments of Inertia for Uniform, Rigid Objects of Various Shapes and Total Mass M
101.	Figure 10-17	An object rolls down an incline
102.	Example 10-6	Spinning Wheel

Physlet® Illustrations

10-1 Angular Velocity
10-2 Angular Acceleration
10-3 Centripetal Force
10-4 Rolling Motion
10-5 Rolling on an Incline

Suggested Readings

Altshuler, K. and Pollock, P., "Inexpensive Rotating-Arm Device for Angular Motion Labs," *The Physics Teacher* (October 1998), pp. 424-425. Describes a very inexpensive device that can be constructed and used for demonstration of angular motion.

Ho, A., Contardi, L., Dion, P. and Griffioen, E., "Rotating Wheels as Seen on Television," *The Physics Teacher* (September 1998), pp. 367-369. An analysis of the motion of rotating wheels as seen under stroboscopic lighting.

Johns, R., "Physics on a Rotating Reference Frame," *The Physics Teacher* (March 1998), pp. 178-180. A description of a simple apparatus useful in illustrating motion in rotating reference frames.

McDonald, K., "Physics in the Laundromat," *American Journal of Physics* (March 1998), pp. 209-211. Analysis of the spin cycle of a washing machine drum.

Mei, W. and Wilkins, D., "Making a Pitch for the Center of Mass and the Moment of Inertia," *The American Journal of Physics* (September 1997), pp. 903-907. Description of some simple experiments to measure moments of inertia of certain physical pendula.

Pritchett, T., Nelson, R., Creamer, T. and Oldaker, B., "Does an Ideal Wheel Really Rotate About Its Instantaneous Point of Contact?," *The Physics Teacher* (March 1998), pp. 167-170. An interesting article on rolling.

Materials

A bicycle wheel gyroscope is available from Fisher Scientific, model number S40965.

A ring and disk suitable for rolling down an inclined plane are also available from Fisher Scientific, model number S41013.

Notes and Ideas

Class time spent on material: Estimated:______ Actual:____________

Related laboratory activities:

Demonstration materials:

Notes for next time:

Chapter 11: Rotational Dynamics and Static Equilibrium

Outline

Summary

Chapter 10 dealt with rotational kinematics. In Chapter 11 we will consider the *causes* of rotational motion. Just as forces cause translational motion, torques cause rotational motion. Angular momentum and rotational work are defined, and the conservation of angular momentum in the absence of external torques is discussed. The chapter ends with an optional section on vectors in rotational motion.

Major Concepts

By the end of the chapter, students should understand each of the following and be able to demonstrate their understanding in problem applications as well as in conceptual situations.

- Torque
  - Definitions
  - Static equilibrium
  - Dynamic applications
- Angular momentum
  - Definitions
  - Conservation of angular momentum
- Rotational work
- Vectors in rotational motion

Teaching Suggestions and Demonstrations

Many students find the concepts of **torque** and **angular momentum** confusing at best. It is important to give many examples (as always!) and to check on the students' progress at regular intervals.

Sections 11-1 and 11-2

All students have experienced **torque**. Talk about the many everyday examples. Go over Figure 11-1 and also discuss the torque involved in opening a standard door, turning a doorknob, and using a lug wrench or screwdriver. For the health-profession majors, you can point out that special doorknobs and water faucets made for people with arthritis are designed to increase r, decrease F, and allow straight-line (rather than turning) motions. You will need to spend some time on the general definition of torque and on the **sign conventions** in Section 11-1.

➲ **DEMO 11-1** Bring in several tools, especially various kinds of screwdrivers and wrenches. Ask the students to decide which ones would provide more torque for the same force. (Ask them to justify their answers.)

Torques cause **angular accelerations**. The equation that connects torque and angular acceleration is $\tau = I\alpha$, where I is the **moment of inertia**. Remind the students that in rotational motion I plays the role that m plays in translational motion. (See the table in Section 11-2 for the other correspondences.) Conceptual Checkpoint 11-1 is an excellent thought experiment.

➲ **DEMO 11-2** Do Example 11-3 (Drop It). It is very clear and requires no special equipment. It is also useful to bring a meter stick to class for this section. You can let it rotate about the center or about an end, and easily measure the distance from the axis of rotation to the line of action of the force.

Section 11-3

Be sure to give several examples of objects in **static equilibrium**, such as bridges, buildings, playground structures, or sawhorses. Point out in each case that just having the **sum of the forces** equal to zero is not enough. The **sum of the torques** must also be zero. You may want to mention that the text restricts consideration to planar forces (x and y directions), which restricts the torques to a single axis.

Emphasize that the **torque is always taken about an axis of rotation**. The axis can be placed at any location, but once placed, it must stay put for the rest of the problem. All torques in the problem must be computed about the same axis. Computations in problems can be made harder or easier by the choice of axis. It is important to choose wisely. Be sure to work several examples. Active Example 11-3 is a good one to work, as it will allow you to talk about the role of the normal force and friction in static equilibrium.

Sections 11-4 and 11-5

The idea that an object **balances** if it is suspended from its center of mass is easy to understand when the object is symmetric, like a meter stick or a barbell. It is much more difficult to understand when the object has a non-uniform mass distribution.

➲ **DEMO 11-3** Bring in a mobile. You can make one out of coat hanger and cardboard and have the students help balance it. Ask them how they know how to balance the system. Once they have figured out what they are doing intuitively, you can quantify their reasoning with the torque equation for static equilibrium and the center-of-mass equation.

Go over Conceptual Checkpoint 11-2, which the students may find confusing. Remind them that the center of mass is a weighted average of the positions of the masses and not necessarily the geometric center of the mass distribution. The photographs in Section 11-4 are also good for discussion.

Section 11-5 concentrates on the **applications of the torque equations** to Newton's second law problems. The definition of $\alpha = a/R$ serves as a bridge between the force equations and the torque equations. Students often forget about this equation; you will have to remind them frequently.

Sections 11-6 and 11-7

There are several points about **angular momentum** that students will not understand immediately. One is that the angular momentum, like torque, depends on the **choice of axis**. Another is that a particle can have angular momentum about an axis while moving in a straight line. Still another is that a particle moving along a radial path has no angular momentum about an axis through the center of the circle. Conceptual Checkpoint 11-3 goes over several of these points.

Remind the students that the angular momentum plays the role in rotational motion that the linear momentum plays in translational motion. Just like the change in linear momentum with time is equal to the force, the change in angular momentum with time is equal to the torque. This implies that if the net torque is zero, angular momentum is constant in time. There are many good examples of **conservation of angular momentum**. Talk about divers, skaters, and gymnasts. Be sure to point out that in all of these cases it is possible to change the moment of inertia, and thus change the angular velocity, without changing the angular momentum.

➲ **DEMO 11-4** Example 11-10 is a great demonstration of **conservation of angular momentum**. Sit on a rotating stool (or better yet, have a student sit on the stool) while holding dumbbells or hand weights at arm's length. (See Resource Information.) Have a student help you start spinning (slowly!). Once you are in motion, bring the weights in close to your torso. As you reduce your moment of inertia, you will increase your angular velocity.

Try this while holding a bicycle wheel instead of the weights. Hold the wheel by the axle and have a student spin it. As you sit on the stool, you can make yourself rotate by trying to change the position of the axis of the wheel. See if the students can explain why!

Active Example 11-4 and the picture next to it are good for discussion. They help show that conservation of angular momentum is a universal principle of physics.

Rotational collisions are interesting to talk about. Students will generally not be familiar with the turntable example, but you can discuss disc brakes and rotating thrust bearings. Active Example 11-5 is easy to visualize and a good example to work through in class.

Section 11-8

The correspondence between the equation for translational work and the equation for **rotational work** makes this concept clear for most students. You will need to discuss the work-energy theorem again and do several examples. Emphasize the simplification of the calculation if problems are treated with the work-energy theorem rather than with rotational kinematics.

*Section 11-9

This optional section goes over the **right-hand rules for angular velocity, angular momentum and torque**. If you have access to a bicycle wheel (or other disk) you can demonstrate these rules quite easily. Many students find the right-hand rules helpful in visualizing rotational motions. It is a good idea to spend a little time on the qualitative aspects of these concepts. We recommend skipping the section on **gyroscopes** and **precession** if you have limited time.

➲ **DEMO 11-5** Demonstrate the precession of the axis of a gyroscope. While you have the gyroscope out, you can also do other fun tricks, such as having it balance on a tightrope. Use this demo to lead into a discussion of why it is easy to stay upright on a moving bicycle (with turning wheels) but nearly impossible to keep from tipping over if the bicycle is stopped.

Resource Information

Transparencies

103. Figure 11-1	Applying a torque
104. Figure 11-2	Only the tangential component of a force causes a torque
105. Figure 11-3	The moment arm
106. Figure 11-5	Forces required for static equilibrium
107. Figure 11-6	A lamp in static equilibrium
108. Active Example 11-3	Find the Forces
109. Figure 11-8	Equilibrium of a suspended object
110. Figure 11-10	A mass suspended from a pulley
111. Figure 11-11	The angular momentum of circular motion
112. Figure 11-12	The angular momentum of non-tangential motion
113. Figure 11-16	The right-hand rule for angular velocity
114. Figure 11-17	The right-hand rule for torque
115. Figure 11-18	Torque and angular momentum vectors
116. Figure 11-19	The torque exerted on a gyroscope
Figure 11-20	Precession of a gyroscope

Physlet® Illustrations

11-1 Torque and Moment of Inertia
11-2 Conservation of Angular Momentum

Suggested Readings

Bracikowski, C., "Feeling the Forces That Produce Torques," *The Physics Teacher* (January 1998), p. 15. A simple modification to the classic introductory experiment that allows students to experience the forces that produce torques.

Cox, A., "Angular Momentum and Motorcycle Counter-Steering: A Discussion and Demonstration," *American Journal of Physics* (November 1998), pp. 1018-1020. A demonstration of torque and angular momentum.

Cross, R., "The Trajectory of a Ball in Lawn Bowls," *American Journal of Physics* (August 1998), pp. 735-738. A discussion of the trajectory and precession of a weighted ball rolling on a smooth horizontal surface.

Denardo, B., "Demonstration of the Parallel-Axis Theorem," *The Physics Teacher* (January 1998), pp. 56-57. A simple demonstration for the parallel-axis theorem.

Gallant, J., "The Shape of the Eiffel Tower," *American Journal of Physics* (February 2002), pp. 160-162. An interesting discussion of the design of the Eiffel Tower and the balance of the maximum torque created by wind and the torque due to the tower's weight.

Gluck, P., "MBL Experiment in Angular Momentum," *The Physics Teacher* (April 2002), pp. 230-234. Studies the loss and conservation of angular momentum using a small direct current motor as generator.

Henderson, C., "Measuring the Forces Required for Circular Motion," *The Physics Teacher* (February 1998), pp. 118-121. A description of an apparatus that allows students to measure the forces exerted in circular motion.

Mohazzabi, P., "Free Fall and Angular Momentum," *American Journal of Physics* (November 1999), pp. 1017-1020. A discussion of the influence of the Earth's rotation on freely falling objects.

Pechan, M., O'Brien, A. and Burgei, W., "Conservation of Angular Momentum Apparatus Using Magnetic Bearings," *The Physics Teacher* (January 2001), pp. 26-28. A simple laboratory exercise on angular momentum.

Schönhammer, K., "Elementary Theoretical Description of the Heavy Symmetric Top," *American Journal of Physics* (November 1998), pp. 1003-1007. A theoretical description of a symmetric spinning top employing conservation of energy and presented without the use of Lagrangian mechanics.

Materials

A rotating stool is available from Fisher Scientific, model number S51543.

Notes and Ideas

*Class time spent on material: Estimated:*_______ *Actual:*______________

Related laboratory activities:

Demonstration materials:

Notes for next time:

Chapter 12: Gravity

Outline

Summary

This chapter introduces Newton's law of universal gravitation and applies it to spherical objects. Kepler's laws are stated and discussed, as is the general equation for gravitational potential energy and its role in the conservation of energy. The chapter ends with an optional section on tides.

Major Concepts

By the end of the chapter, students should understand each of the following and be able to demonstrate their understanding in problem applications as well as in conceptual situations.

- Newton's law of universal gravitation
  - Universal gravitation constant *G*
  - Inverse square dependence on the distance
  - Point and spherical objects
  - Cavendish experiment
- Kepler's laws of orbital motion
  - Law of orbits
  - Law of areas
  - Law of periods
- Gravitational potential energy

Teaching Suggestions and Demonstrations

All students have experience with the **gravitational force**, but not all of them will connect the force we feel on Earth with the one that holds the solar system (and the galaxy and the universe) together. Some of their intuitive ideas about gravity will be correct; others will not be correct. You will have to make careful assessment of student progress for this chapter.

Sections 12-1 and 12-2

Newton's law of universal gravitation is elegant and simple, especially for spherical or point objects. Emphasize that the force acts between centers of the objects, is always attractive, is proportional to both masses, and is inversely proportional to the distance between the centers squared. You will need to discuss the implications of the inverse square nature of the force. Figure 12-2 will help. The **infinite range of the gravitational force** and the fact that it is the weakest of the **four fundamental forces** are often surprising to students. It is a good idea to talk about why we don't have to worry about the gravitational force from everyday objects. (See Exercise 12-1.) For most cases, the only time we have to take the gravitational force into account is when at least one of the two objects is the size of a planet. The global model of the gravitational field strength of the Earth in Section 12-2 is interesting and a good basis for discussion.

There is a nice treatment of the **Cavendish experiment** in Section 12-2. Point out that ***G*** is a universal constant; it is the same for all pairs of objects. Students often find it amazing that the mass of the Earth wasn't known until 1798, long after Newton's death, since the radius of the Earth was known to the ancient world. The calculation of the average density of the Earth and its implications about the interior structure of the Earth are also interesting.

Section 12-3

Kepler's laws can be derived from Newton's laws, but Kepler deduced them from astronomical data collected by visual observation. The story of Brahe and Kepler makes interesting reading; students find it fascinating. (See Resource Information.)

Kepler's first law is also called the **law of orbits.** Giving up the idea of circular orbits for planets involved a major shift of worldview for Kepler. You will probably need to spend some time talking about the properties of ellipses before continuing to the second law.

➲ **DEMO 12-1** Using suction cups attached to the blackboard or whiteboard and a piece of string tied in a loop, draw an **ellipse**. Show the students what happens if the foci are moved farther apart or closer together. Show them how a circle is a special case of an ellipse.

Kepler's second law is called the **law of areas**. Point out that it is based in the conservation of angular momentum.

Kepler's third law is called the **law of periods**. It is important to emphasize to the students that the "constant" that appears in the third law is *not* the universal constant *G*. It is a "constant" that is the same for all objects orbiting the same mass. Active Example 12-1 steps students through the calculation for the radius of the orbit of **geosynchronous satellites**. Note the discussions of the Global Positioning System (GPS) and orbital maneuvers in this section.

Sections 12-4 and 12-5

In Chapter 8 we defined the **gravitational potential energy near the surface of the Earth**. We now define the **general gravitational potential energy**. You will need to discuss the new **conventions** for the general gravitational potential energy: The zero is set at r = *infinity*, the potential energy is taken to be negative for all values of r smaller than infinity, and r is measured from the center of one object to the center of the other. Go through the calculation in Section 12-4 that connects the two expressions for gravitational potential energy.

➲ **DEMO 12-2** Make a model of a **gravitational potential well**. Stretch a balloon (cut off the end) over a cookie tin. Place a lead weight or another heavy object in the middle. The weight will deform the balloon. If you place a marble at the edge of the cookie tin, it will roll toward the middle. (You can even discuss the curvature of space-time if you like!)

Since the gravitational force is a conservative force, we can write a **conservation of energy** expression that includes the general form of the gravitational potential energy. The calculation of the speed of an asteroid or comet just before it strikes the Earth in Section 12-5 is fascinating. Be sure to go through the calculation of **escape speed** and discuss how it impacts planetary atmospheres. The material on **black holes** is also interesting but can be skipped if necessary.

*Section 12-6

Make this optional section on **tides** a reading assignment. Talk briefly and qualitatively about Figure 12-19. Ask the students why there are two high tides and two low tides a day.

Resource Information

Transparencies

117. Figure 12-2	Dependence of the gravitational force on separation distance, *r*
118. Figure 12-3	Gravitational force between a point mass and a sphere
119. Figure 12-4	The acceleration due to gravity at a height *h* above the Earth's surface
120. Figure 12-6	The Cavendish experiment
121. Figure 12-8	The circle as a special case of the ellipse
122. Figure 12-9	Kepler's second law
123. Figure 12-10	Kepler's third law and some near misses
124. Figure 12-12	The Global Positioning System
125. Figure 12-13	Orbital maneuvers
126. Figure 12-15	Potential and kinetic energies of an object falling toward Earth
127. Figure 12-16	A gravitational potential "well"
128. Figure 12-19	Tides

Physlet® Illustrations

12-1 Kepler's Laws

Suggested Readings

Adam, D., "Amazing Grace," *Nature* (7 March 2002), v. 416, pp. 10-11. A very accessible article describing the precise mapping of Earth's gravitational field by two satellites.

Metz, J., "Finding Kepler's Third Law with a Graphing Calculator," *The Physics Teacher* (April 2000), p. 242. Instructions for using a TI-83 to find the relationship between orbital radius and period for planets.

Morris, R., *Dismantling the Universe: The Nature of Scientific Discovery*, Simon and Schuster, New York, 1983. A great book to read before teaching a physics class. Chapter 4, pp. 81-101, covers the story of Brahe and Kepler (and Galileo).

Sawicki, M., "Myths About Gravity and Tides," *The Physics Teacher* (October 1999), pp. 438-441. Explores popular misconceptions about gravity and tides.

Schwarzschild, B., "Theorists and Experimenters Seek To Learn Why Gravity Is So Weak," *Physics Today* (September 2000), pp. 22-24. A description of the search for departures from the inverse-square law at millimeter separations.

Toepker, T., "Babies and the Moon," *The Physics Teacher* (April 2000), p. 242. A graph of birth data to dispel the popular myth that more babies are born under a full moon.

Vogt, E., "Elementary Derivation of Kepler's Laws," *American Journal of Physics* (April 1996), pp. 392-396. A proof of Kepler's laws that follows from conservation of energy and angular momentum, with further discussion.

Wright, K., "Very Dark Energy," *Discover* (March 2001), pp. 70-76. A discussion of the new ideas on the accelerating expansion of the universe.

Notes and Ideas

*Class time spent on material: Estimated:*______ *Actual:*____________

Related laboratory activities:

Demonstration materials:

Notes for next time:

Chapter 13: Oscillations About Equilibrium

Outline

13-1 Periodic Motion
13-2 Simple Harmonic Motion
13-3 Connections Between Uniform Circular Motion and Simple Harmonic Motion
13-4 The Period of a Mass on a Spring
13-5 Energy Conservation in Oscillatory Motion
13-6 The Pendulum
13-7 Damped Oscillations
13-8 Driven Oscillations and Resonance

Summary

General periodic motion and the special case of simple harmonic motion are treated thoroughly in Chapter 13. Students use their knowledge of circular motion to derive equations for the position, velocity and acceleration of an object undergoing simple harmonic motion. The specific cases of a mass on a spring and a pendulum are emphasized. Conservation of energy is applied to oscillating systems. Finally, real-world situations involving physical pendulums, damped oscillations and resonance are discussed.

Major Concepts

By the end of the chapter, students should understand each of the following and be able to demonstrate their understanding in problem applications as well as in conceptual situations.

- Periodic motion
  - Frequency
  - Period
- Simple harmonic motion
  - Sine and cosine curves
  - Connection to uniform circular motion
  - Position, velocity, acceleration
  - Angular frequency
- Mass on a spring
- Simple pendulum
- Conservation of energy applied to oscillating systems
- Damped and driven oscillations and resonance

Teaching Suggestions and Demonstrations

Until this point in the course, most of the trigonometry that students have been using has involved the basic sine, cosine, and tangent function definitions applied to triangles. Now is the time to remind students to review the **graphs of sine and cosine functions** or even review briefly in class if you have time. In order to tackle oscillatory motion, students need to be comfortable expressing angles in radians and sketching graphs of sine and cosine functions.

Sections 13-1 through 13-3

Begin by asking students to name as many examples of **periodic motion** as they can. The heartbeat pictured in Section 13-1 is a common one. You can also bring in a graph of tides to illustrate the wide variety in motions that are periodic. Next, students can derive Equation 13-1, the relationship between **frequency** and **period**. They will understand the inverse relationship better if they think about it before they actually see the equation. For instance, ask them, "If you run around a track six times in one hour, how long did it take you to complete just one lap?" Six cycles per hour is the frequency and one sixth of an hour (or ten minutes) is the period. Try other examples with the more common units of **hertz** (cycles per second) for the frequency and seconds for the period.

➲ **DEMO 13-1** Throughout your discussion of this chapter, it's a good idea to have a **spring** and a **simple pendulum** on hand. Set up an oscillation with either and have students determine frequency and period and verify the relationship between them. Some students will probably have stopwatches built into their wristwatches. Others can try Galileo's technique (see the Real World Physics discussion in Section 13-6) and use their pulse rate as a timing device!

The spring and pendulum lead the discussion nicely from general periodic motion to **simple harmonic motion**. Derive the equations for the position, velocity and acceleration of an object in simple harmonic motion by using the connection to **uniform circular motion**, as shown in Section 13-3. Students will need to spend some time with these equations after they have been derived to fully understand them. Superimposing all three graphs can help emphasize the phase relationships. To tie it in to real life motion, set the spring oscillating again and ask them to describe the relationship between velocity and position. Conceptual questions are also helpful. If the position of the mass is positive and the velocity is negative, where might the mass be? Which direction is it moving? Is it moving toward or away from its equilibrium position? Is it speeding up or slowing down? Is its acceleration positive or negative? What point on the graph might represent this point in the oscillation?

Often, students have the most trouble conceptualizing the **acceleration graph**. Remind them of Newton's second law: The acceleration is proportional to the force exerted by the spring. From Hooke's law, introduced in Chapter 6, we also know that the force is a restoring force, proportional to the displacement but in the opposite direction. Relating both position and acceleration directly to force in this way clarifies the relationship between them.

Sections 13-4 through 13-6

➲ **DEMO 13-2** Before deriving the equation for the **period of a mass on a spring**, bring in a number of different springs and masses and lead students in a determination of which factors influence period and which do not. Does increasing the mass increase, decrease, or have no effect on the period? Does the same mass on a stiffer spring have a period that is longer, shorter, or the same? Does increasing the amplitude (that is, how far the mass is pulled from equilibrium before it is released) have any effect?

The demonstration will lead students to qualitative results, such as the fact that when mass increases, so does period. The exact dependence—period increases as the square root of mass—will become clear in the derivation. Equating Hooke's law and Newton's second law leads to an expression for the **angular frequency for a mass on a spring** in terms of the mass and the spring constant. The period is defined from the angular frequency. Make sure students understand the graphs in Figure 13-8, and give them a chance to generate some for similar situations, such as increasing k or m by a factor of 9. The Real World

Physics discussion on NASA's Body Mass Measurement Device gives an interesting example of an application of the relationship between mass and period.

➲ DEMO 13-3 The **period of a simple pendulum** can be introduced the same way as the period of a mass on a spring, by letting students adjust different variables and determine which ones affect the period and how. It will probably come as a surprise to them that changing the mass does not change the period. (Remind them that if air resistance can be neglected, all objects near the surface of the Earth accelerate at the same rate.)

In deriving the equation for the **period of a simple pendulum**, refer to Figure 13-15 to help students understand the small angle approximation and that the equation is valid only for oscillations with small amplitudes. You may not have enough time to discuss the case of the physical pendulum in detail, but it is worth pointing out that legs are essentially pendulums, though not simple ones. Ask students if they have seen Dachshunds walk. The period of oscillation for a Dachshund leg is very different from that of a Great Dane!

Another interesting application is the **Foucault pendulum**, invented in 1851 and used to demonstrate the Earth's rotation. The pendulum appears to change its path throughout the course of the day, but in fact the floor under it is moving, carried by the rotation of the Earth.

Potential energy of a spring was introduced in Chapter 8. The graph in Figure 13-10 illustrates that the **total energy of a mass on a spring** remains constant (neglecting friction) while the proportions of kinetic and potential vary. Example 13-6 or one like it is a good one to start with. Adding friction between the block and the table is a nice extension and provides students with a review of friction and of the usefulness of energy considerations even when mechanical energy is not conserved. Continue with Active Example 13-3, which requires a consideration of both energy and linear momentum. You can also do a ballistic pendulum example, in which a bullet embeds in a pendulum, raising it a certain height. The problem requires consideration of both conservation of linear momentum during the bullet/pendulum collision and conservation of energy as kinetic energy is converted to potential energy.

➲ DEMO 13-4 A Newtonian collision ball apparatus, like that shown in the photo in Section 9-6, is a fairly easy and inexpensive way to demonstrate pendulum motion as well as conservation of energy and conservation of linear momentum. (See Resource Information.)

Sections 13-7 and 13-8

Resonance and natural frequency are important concepts to touch on here because both appear in the course later with waves and again with AC circuits. The photos at the end of the chapter show two good examples. Most students will be familiar with the "feel" of pushing someone on a swing; they just won't realize that what they are doing is driving a system at its resonant frequency! The video of the collapse of the Tacoma Narrows bridge is quite impressive and well worth showing to your class. (See Resource Information.)

Resource Information

Transparencies

129. Figure 13-1	A mass attached to a spring undergoes simple harmonic motion about $x = 0$
130. Figure 13-2	Displaying position versus time for simple harmonic motion

131. Figure 13-3 Simple harmonic motion as a sine or a cosine
132. Figure 13-5 Position versus time in simple harmonic motion
133. Figure 13-6 Velocity versus time in simple harmonic motion
134. Figure 13-7 Acceleration versus time in simple harmonic motion
135. Figure 13-8 Factors affecting the motion of a mass on a spring
136. Figure 13-10 Energy as a function of position in simple harmonic motion
137. Figure 13-11 Energy as a function of time in simple harmonic motion
138. Figure 13-12 Motion of a pendulum
139. Figure 13-13 The potential energy of a simple pendulum
140. Figure 13-14 Forces acting on a pendulum bob
141. Figure 13-15 Relationship between sin θ and θ
142. Figure 13-18 Damped oscillations
143. Figure 13-20 Resonance curves for various amounts of damping

Physlet® Illustrations

13-1 Simple Harmonic Motion
13-2 Energy and the Mass on a Spring

Suggested Readings

Bensky, T., "Measuring *g* with a Joystick Pendulum," *The Physics Teacher* (February 2001), pp. 88-89. Instructions for making a physical pendulum to measure *g*.

Erlichson. H., "Galileo's Pendulum," *The Physics Teacher* (November 1999), pp. 478-479. Description of a lab experiment investigating the small angle approximation and discussion of Galileo's assertion that the period of a pendulum is independent of the amplitude.

Greene, N. and Dunn, R., "A Conical Spring - Which End Up?," *The Physics Teacher* (April 2000), pp. 228-231. Useful information about the conical spring commonly used in physics labs.

Holzwarth, D. and Malone, J., "Pendulum Period Versus Hanging-Spring Period," *The Physics Teacher* (January 2000), p. 47. A student exercise in matching periods of oscillating masses.

Kidd, R. and Fogg, S., "A Simple Formula for the Large-Angle Pendulum Period," *The Physics Teacher* (February 2002), pp. 81-83. Extends the theory and experiments typical of introductory physics discussions to include large-amplitude pendulum oscillations.

Peters, R., "Student-Friendly Precision Pendulum," *The Physics Teacher* (October 1999), pp. 390-393. Instructions for constructing a Kater pendulum to accurately measure *g*.

Szapiro, B., "Simple-Pendulum Lab with a Twist," *The Physics Teacher* (March 2002), pp. 158-163. Uses a force sensor connected to a computer to overcome some of the difficulties with a standard pendulum lab.

Materials

AAPT, "Twin Views of the Tacoma Narrows Bridge Collapse," video, user guide, and student activities. Order online at: www.aapt.org/catalog.

A Newtonian collision ball apparatus is available from Fisher Scientific, model number S40976.

Drilled balls for use as pendulum bobs are available from Frey Scientific.

A Hooke's law spring set is available from Sargent-Welch, model number CP33855-00.

Notes and Ideas

Class time spent on material: Estimated:_________ Actual:______________

Related laboratory activities:

Demonstration materials:

Notes for next time:

Chapter 14: Waves and Sound

Outline

14-1 Types of Waves
14-2 Waves on a String
*14-3 Harmonic Wave Functions
14-4 Sound Waves
14-5 Sound Intensity
14-6 The Doppler Effect
14-7 Superposition and Interference
14-8 Standing Waves
14-9 Beats

Summary

The first three sections of Chapter 14 introduce the student to basic wave properties and characteristics. Sound waves are discussed next, in Sections 14-4 through 14-6. The final three sections of the chapter deal with superposition and interference; standing waves, which are created when a wave interferes with its reflection; and beats, which result from the interference of two waves with slightly different frequencies.

Major Concepts

By the end of the chapter, students should understand each of the following and be able to demonstrate their understanding in problem applications as well as in conceptual situations.

- Waves
  - Transverse and longitudinal
  - Wavelength and frequency
  - Speed of a wave
- Sound waves
  - Speed of sound
  - Frequency and pitch
  - Intensity and intensity level
  - The Doppler effect
- Superposition and interference
  - Constructive and destructive
  - Phase
  - Standing waves (strings and pipes)
  - Beats

Teaching Suggestions and Demonstrations

Now is a good time to find out if you have any musicians or music majors in your class. If so, this is the chapter for them! Elicit their help when you get to the standing waves portion of the chapter.

Sections 14-1 through 14-3

Basic definitions and **wave characteristics** are covered in the first section of this chapter. Emphasize Equation 14-1, relating **speed, wavelength and frequency**, which will reappear when light is discussed later in the course.

➲ **DEMO 14-1** A Slinky® is a handy prop to bring to class throughout this chapter. (See Resource Information.) As your first demonstration, use it to illustrate the difference between **transverse and longitudinal waves**. Transverse waves are set up by shaking one end of the stretched Slinky® up and down, longitudinal by gathering a number of coils in your hand and releasing them. Tie a piece of red yarn somewhere on the Slinky® and repeat the demonstrations a few times so that students can watch the whole Slinky® and then concentrate on just the motion of the yarn. The yarn not only emphasizes the difference between the types of waves, since it moves perpendicular to the direction of propagation of the wave pulse in one case and parallel in the other, it also illustrates the basic fact that a wave is the **propagation of energy**, not of matter. Even though the pulse makes it to the opposite end of the Slinky®, the yarn never moves far from its original position.

➲ **DEMO 14-2** A climbing rope or long jump rope works well for investigating the **speed of a wave on a string**. Fix one end of the rope and send a pulse along its length; repeat for different tensions. It should be apparent that speed increases if tension increases. The square root dependence will not be obvious from this qualitative demonstration, but the dimensional analysis shown in Section 14-2, while not a derivation, makes a convincing argument. The rope can also be used to demonstrate **reflections**. The behavior of a pulse when it reaches the end of the rope depends on whether or not the end is fixed. A similar phenomenon for light waves reflecting off boundaries between different media will be very important to the thin film interference discussion in Chapter 28.

The typical "picture" of a **harmonic wave** is a sine or cosine curve. It is worth spending a little time ensuring that students know what they are looking at. Point out that you can't really have a complete **representation of a wave with a graph** because a wave is moving. What you can have is either a snapshot of the whole wave at one instant in time (y *vs.* x) or a graph of the displacement of one bit of medium over time (y *vs.* t). In the Slinky® demonstration, the "snapshot" would be a picture of the whole Slinky® at one moment, and the "graph" would be the position of the red yarn as a function of time.

Sections 14-4 through 14-6

Most students will probably be familiar with the rule of thumb that relates the time elapsed between seeing a lightning flash and hearing a thunder clap to the distance to the lightning strike. Use Conceptual Checkpoint 14-2 to illustrate the **speed of sound** and explain why the rule of thumb works. You can point out that light travels at a finite speed too, but that the speed is so great (3×10^8 m/s or 186,000 miles/s) that the light from a source as close as a lightning bolt arrives practically instantaneously.

You won't need to spend much class time on **ultrasonic and infrasonic sound**, although students should be aware that these are nothing more than sounds with frequencies beyond the human range of perception. The photographs and Real World Physics discussions at the end of Section 14-4 are great examples. If you have students in the pre-health professions in your class, be sure to point out the many applications of ultrasound to the medical field.

The difference between **sound intensity and intensity level** tends to be a hard concept for students because of the logarithmic nature of the decibel scale. Convince them that the use of the **logarithmic scale** makes sense for two reasons: first, the range of sound intensities we can hear is very large (refer to Table 14-2), and second, our ears perceive sound logarithmically. That is, if we hear one sound to be twice as loud as another, it has an intensity (energy per time per area) that is 10 times as great. Write Table 14-2 on the board or an overhead and add an extra column. Then have students, working alone or in small groups, find the intensity level (in decibels) that corresponds to each of the intensities listed and record them on the table. This exercise serves two purposes: It gives each student a chance to practice with the equation, and it also produces a list of intensity levels for familiar sounds. Do Example 14-4 or another one like it that you make up to further illustrate the nature of a logarithmic scale. It makes perfect sense to students that adding a second, identical crying baby to the first doubles the sound intensity. However, it only increases the intensity *level* from 69 dB to 72 dB.

To introduce the **Doppler effect**, find a student to imitate the sound of a siren coming toward, passing and then speeding away from a stationary car. As an analogy, have them imagine a person in a boat in the middle of a pond dropping pebbles into the water at a constant frequency. The frequency and wavelength of the ripples that reach an observer on the shore will depend on whether the rowboat is stationary, moving toward the shore, or moving away from the shore.

➲ **DEMO 14-3** A Nerf® ball with a beeper or bell embedded inside it makes a nice **Doppler ball**. Students throw the ball back and forth while the class listens to the sound and compares it to the sound they hear when the ball is stationary. (See Resource Information.)

Sections 14-7 through 14-9

To introduce students to **superposition and interference** of sound waves, have them consider interfering water waves first. They will probably be familiar with the patterns that arise from interfering circular waves created when two pebbles are dropped in water. Figure 14-21 is a good pictorial representation. It will likely be harder for them to understand that two sounds can interfere and cancel each other out, but reminders of the wave nature of sound help. Refer also to Figure 14-20. If possible, actually set up speakers and demonstrate the situation shown in Example 14-7 and Active Example 14-2.

Introduce **standing waves** as an interference of a wave with its reflection under the right conditions. Drawing the two interfering waves at quarter-period intervals helps show students that the interference will alternate between constructive and destructive. Even when the interference is constructive, though, the nodes remain still.

➲ **DEMO 14-4** A Slinky® or a long spring can be used to demonstrate **standing waves**. If students are doing the demonstration, challenge them to create standing waves representing the different harmonics in order, with an increasing number of nodes and antinodes. In addition to seeing the standing waves, they will recognize that they need to increase the frequency of vibration to create more antinodes. Relate this observation to the relevant formulas. In this case, the length and the tension are kept constant.

➲ **DEMO 14-5** You can also demonstrate **standing waves** with a string attached to a vibrator. (See Resource Information.) Orient the string horizontally with the free end over a pulley and attached to a weight hanger. As different amounts of weight are added to the vibrating string, different standing wave patterns will be produced. In this case, frequency and length are kept constant, but the tension in the string, provided by the weights, is a variable. Notice that a higher tension is required to produce fewer antinodes. Again, have students relate the observations to the relevant equations. This demonstration can also be done quantitatively. Measure the mass and length of the string as well as the force on the end, and the frequency can be calculated and compared to the known vibrator frequency.

➲ DEMO 14-6 **Standing waves in a column of air** can be demonstrated by blowing across the top of a bottle as in Figure 14-26 and Example 14-9. Better yet, use a resonance tube so that the water level, and therefore the length of the air column, can be adjusted. Strike a tuning fork at the top of the tube and quickly raise or lower the water level until resonance is heard. (See Resource Information.)

➲ DEMO 14-7 To illustrate **beats**, use two tuning forks with frequencies close to each other or try a demonstration of Example 14-10. You can also use two identical tuning forks and wrap a rubber band around one tine of one fork. This reduces the frequency slightly, producing beats when both forks are struck.

The strings on **instruments** like guitars and violins oscillate in standing wave patterns to produce notes. Considering such instruments helps students review many of the concepts and formulas from the chapter. For instance, most students will know that when you tighten a string, the pitch increases. Tightening corresponds to increasing the tension, which will increase the speed of waves on the string (Equation 14-2). Increasing the speed while keeping the wavelength (or the length of the string) constant increases the frequency (Equations 14-1 and 14-12), which is related to pitch (see Section 14.4). You can also examine what happens to the pitch when a thicker string is used, which changes the mass per unit length, or when the musician uses a capo, which effectively changes the length of the string.

Resource Information

Transparencies

144. Figure 14-2	The motion of a wave on a string
145. Figure 14-3	Sound produced by a speaker
146. Figure 14-6	The speed of a wave
147. Figure 14-7	A reflected wave pulse: fixed end
148. Figure 14-8	A reflected wave pulse: free end
149. Figure 14-9	A harmonic wave moving to the right
150. Figure 14-10	A wave on a Slinky
Figure 14-11	Wave properties of sound
151. Figure 14-12	Intensity of a wave
152. Table 14-2	Sound intensities (W/m²)
153. Figure 14-15	The Doppler effect: A moving observer
154. Figure 14-16	The Doppler effect: A moving source
Figure 14-17	The Doppler-shifted wavelength
155. Figure 14-20	Interference
156. Figure 14-21	Interference of circular wave
157. Figure 14-23	A standing wave
158. Figure 14-24	Harmonics
159. Figure 14-27	Standing waves in a pipe that is open at one end
160. Figure 14-29	Standing waves in a pipe that is open at both ends

161. Figure 14-30 Interference of two waves with slightly different frequencies
Figure 14-31 Beats

Physlet® Illustrations

14-1 Introduction to Waves
14-2 Doppler Effect
14-3 Standing Waves in a Pipe
14-4 Beats

Suggested Readings

Bedard, A. and Georges, T., "Atmospheric Infrasound," *Physics Today* (March 2000), pp. 32-37. Discussion of detection of sounds with frequencies too low for humans to hear, and various applications including monitoring compliance with the Comprehensive Test Ban Treaty.

Carnevali, A. and Newton, C., "Coupled Harmonic Oscillators Made Easy," *The Physics Teacher* (November 2000), pp. 503-505. An experiment for analyzing the frequencies of coupled harmonic oscillators.

Dibble, W., "A Pedagogical Note on the Doppler-Effect Formulas," *The Physics Teacher* (September 2000), pp. 362-363. A quick and simple derivation for the Doppler effect formulas useful at the introductory level.

Gibson, G. and Johnston, I., "New Themes and Audiences for the Physics of Music," *Physics Today* (January 2002), pp. 42-48. Discussion of the treatment of music in physics class.

Greenslade, T., "Models of Traveling Waves," *The Physics Teacher* (November 2001), p. 466. Suggestion for how to make a three dimensional model of a traveling wave, in order to help students understand that y is a function of both x and t.

Greenslade, T., "Waves in the Movies," *The Physics Teacher* (February 2000), p. 78. Nice demonstration of transverse waves.

Haar, G., "Acoustic Surgery," *Physics Today* (December 2001), pp. 29-34. Describes the focusing of high intensity ultrasound fields for medical procedures, including the treatment of cancer.

Hartmann, W., "How We Localize Sound," *Physics Today* (November 1999), pp. 24-29. Describes a variety of cues our brain uses to construct a three-dimensional image of the acoustic landscape.

Hoyt, D., "A Different Viewpoint on Doppler-Effect Calculations," *The Physics Teacher* (January 2002), pp. 14-16. Analysis of the classical Doppler-effect formula, including a method for determining signs and a useful approximation.

Kanamori, H. and Brodsky, E., "The Physics of Earthquakes," *Physics Today* (June 2001), pp. 34-40. A discussion of information gleaned from seismic waves as well as many other applications of physics to earthquakes.

Krider, J., Cook, T., Easterling, W. and Rzanek, I., "Hear and 'See' Sound Interference," *The Physics Teacher* (April 2001), pp. 220-222. A visual demonstration of wave interference.

Martin, B., "Measuring the Speed of Sound—Variation on a Familiar Theme," *The Physics Teacher* (October 2001), pp. 424-426. A new twist on the classic resonance tube lab measurement of the speed of sound.

Pettersen, I., "Speed of Sound in Gases Using an Ultrasound Motion Detector," *The Physics Teacher* (May 2002), pp. 284-286. Experiments to determine the speed of sound in air and other gases, with suggestions for possible areas of further investigation.

Rowland, D. and Pask, C., "The Missing Wave Momentum Mystery," *American Journal of Physics* (May 1999), pp. 378-388. Discussion of paradoxes involved in the study of momentum carried by transverse waves on a string.

Saba, M. and Rosa, R., "A Quantitative Demonstration of the Doppler Effect," *The Physics Teacher* (October 2001), pp. 431-433. Uses a recording of the sound of a passing car horn and spectrogram software to quantitatively investigate the Doppler effect.

Sanchez-Lavega, A., "Acoustic Waves in Planetary Atmospheres," *The Physics Teacher* (April 2002), pp. 239-242. An interesting discussion of sound on other planets and moons in our solar system.

Slogoff, H. and Berner, B., "A Simulation of the Tacoma Narrows Bridge Oscillations," *The Physics Teacher* (October 2000), pp. 442-443. A demonstration of bridge oscillations using a fan and a box with detailed instructions.

Materials

AAPT, "Twin Views of the Tacoma Narrows Bridge Collapse," video, user guide, and student activities. Order online at: www.aapt.org/catalog.

A super Slinky® is available from Fisher Scientific, model number S42162.

One example of a "Doppler ball" is the Vortex Mega Howler, available from Toys R Us or other toy stores.

A resonance apparatus and tuning forks are available from Fisher Scientific, model numbers S42105 and S42036.

A string vibrator is available from Sargent-Welch, model number CP85102-00.

Notes and Ideas

Class time spent on material: Estimated:______ Actual:_____________

Related laboratory activities:

Demonstration materials:

Notes for next time:

Chapter 15: Fluids

Outline

15-1 Density
15-2 Pressure
15-3 Static Equilibrium in Fluids: Pressure and Depth
15-4 Archimedes' Principle and Buoyancy
15-5 Applications of Archimedes' Principle
15-6 Fluid Flow and Continuity
15-7 Bernoulli's Equation
15-8 Applications of Bernoulli's Equation
*15-9 Viscosity and Surface Tension

Summary

Chapter 15 begins with general definitions of density and pressure and then moves to a study of fluids, substances that readily flow. Both liquids and gases are fluids. Fluid statics, the study of fluids at rest, and fluid dynamics, the study of moving fluids, are both examined in this chapter.

Major Concepts

By the end of the chapter, students should understand each of the following and be able to demonstrate their understanding in problem applications as well as in conceptual situations.

- Density
- Pressure
  - General definition
  - Pressure exerted by a fluid
  - Atmospheric, gauge, and actual pressure
- Fluid statics
  - Equilibrium
  - Pascal's principle
  - Archimedes' principle and buoyancy
- Fluid dynamics
  - Equation of continuity
  - Bernoulli's equation

Teaching Suggestions and Demonstrations

There are many everyday applications of the concepts in Chapter 15. Students will know that airplanes can fly and boats can float. They've put air in their car tires and heard that the barometer is falling. And if their dance partner is going to step on their toes, they would rather it be with flat-bottomed shoes than with high heels! In this chapter, you will help them relate all these experiences to physics.

Sections 15-1 and 15-2

Before beginning a study of fluids, students need to understand **density** and **pressure**. Make sure some of your examples involve solids so that students realize that these general definitions apply to substances that aren't fluids as well as to liquids and gases. Emphasize the difference between **gauge pressure** and **actual pressure**. A completely flat tire has a gauge pressure of zero but an actual pressure of about 14.7 psi, since the pressure in the tire equals the atmospheric pressure.

➲ **DEMO 15-1** An assortment of different blocks, some the same size but made out of different materials, others of the same material but different sizes, can help you check student understanding of **density**. Hold up pairs of blocks and ask students which has the greater density. The key is to pay attention to the material, not the size. You can also preview Archimedes' principle by asking them to guess which blocks would float in water and which would sink.

➲ **DEMO 15-2** An assortment of shoes makes a nice demonstration of **pressure**. A person with a given weight will exert different pressures on the floor depending on the surface area of the shoes in contact with the floor. You can do rough calculations by measuring the bottoms of the shoes. On a hot summer day, high heels can actually sink into asphalt parking lots! Winter backpackers wear snowshoes or cross country skis to keep from sinking into the snow. Ice skates work because the thin blade creates so much pressure that the ice right under the blade melts.

Sections 15-3 through 15-5

Students will intuitively understand that **pressure in a liquid** increases with depth, but you may need to work harder to convince them that only the depth matters. If a funnel-shaped container and a cylindrical container are filled to the same height with water, the pressure at the bottom of each will be the same. Students will be tempted to think that the funnel will somehow "concentrate" the pressure. Figure 15-4 and the discussion with it will also help clarify the relationship between depth and pressure. Point out that the formula doesn't work for atmospheric pressure since the density of the atmosphere is not constant. A barometer is a clever device that balances atmospheric pressure with pressure due to an incompressible fluid so that the formula can be applied. The common unit of atmospheric pressure, **millimeters of mercury**, is a direct result of this measuring device. Have students calculate what atmospheric pressure in millimeters of water would be and they will understand why using a denser liquid like mercury is more practical.

While on the topic of barometers, you can take a few minutes to discuss other **weather topics** related to pressure. Point out that fair weather is associated with high-pressure systems and stormy or rainy weather with low. Hurricanes and tornadoes have very low pressure at their centers. When a hurricane passes over a house, the outside pressure drops suddenly, and if the inside pressure doesn't drop also, the roof may get blown off. Students living in hurricane-prone areas will have heard that it is a good idea to open windows if a hurricane is on the way.

The hydraulic lift shown in Figure 15-6 is a great example of **Pascal's principle**. Remind students that energy is still conserved, so the magnification of the force has to be "paid for" somehow. The distance through which the force is applied is the key. A small force applied over a large distance results in the same work as a large force applied over a small distance.

➲ **DEMO 15-3** Introduce **Archimedes' principle and buoyancy** through a demonstration. Suspend a metal block from a spring scale to find its weight. Then lower the block into a beaker of water and notice that the force registered by the spring scale is less. Draw a force diagram to illustrate that another force is acting upward on the block when it is submerged in the water. This is the **buoyant force**, equal to the weight of the displaced water.

After deriving Archimedes' principle, do lots of examples with both **submerged and floating objects**. Students will quickly notice that objects with densities less than water's will float in water, and objects with densities greater will sink. They will tend to think that the buoyant force on an object depends on its density, which isn't quite true. If blocks with different densities but equal volumes are submerged, they will all have the same buoyant force since they all displace the same volume of water. Also do some examples for objects floating in liquids other than water. Conceptual Checkpoints 15-5 and 15-6 provide a good check of student understanding.

Sections 15-6 through 15-8

The final sections of the chapter deal with fluids in motion. The **equation of continuity** will probably be intuitive for students. They will be familiar with the fact that when the cross-sectional area of a pipe decreases, the fluid flowing through it speeds up. That's why you put your thumb over the end of a hose when you want the water to spray, as shown in the photograph in Section 15-6. However, **Bernoulli's equation** may actually be counterintuitive. In the case where potential energy remains constant, the pressure in a flowing fluid will be less where the speed is greatest.

➲ **DEMO 15-4** Begin your demonstrations of **Bernoulli's equation** by blowing across the top of a piece of paper, as shown in Figure 15-16. For a variation, make a paper arch out of the paper and set it on a tabletop. Blow through the arch. Instead of blowing the arch off the table, your breath flattens it! For another dramatic demonstration, place a Ping Pong ball in a funnel and challenge a student to blow it out. When air is blown through the funnel, the air-pressure difference between the top and bottom of the Ping Pong ball will hold the ball in place, even when the funnel is inverted.

Two more examples to discuss are airplane wings (see Figure 15-17) and curve balls in baseball. The velocity of the air on either side of a spinning ball is different because of the drag caused by the ball itself. The side with higher speed air will have lower pressure, so the ball will curve unexpectedly in that direction. **Torricelli's law** is a good example to include in your discussion as well because it illustrates an application involving change in potential energy.

*Section 15-9

This optional section discusses viscosity and surface tension. These topics are interesting but can be skipped without loss of continuity.

Resource Information

Transparencies

162. Table 15-1	Densities of Common Substances
163. Figure 15-3	A simple barometer
164. Figure 15-4	Fluids seek their own level
165. Example 15-4	Oil and Water Don't Mix
166. Figure 15-6	A hydraulic lift
167. Figure 15-7	Buoyant force due to a fluid

168. Figure 15-8	Buoyant force equals the weight of displaced fluid
169. Figure 15-9	Flotation
170. Figure 15-10	Floating an object that is more dense than water
171. Figure 15-13	Fluid flow through a pipe of varying diameter
172. Figure 15-14	Work done on a fluid element
173. Figure 15-15	Fluid pressure in a pipe of varying elevation
174. Figure 15-17	Airflow and lift in an airplane wing
175. Figure 15-20	An atomizer
176. Figure 15-21	Fluid emerging from a hole in a container
Figure 15-22	Maximum height of a stream of water
177. Figure 15-23	Fluid flow through a tube
178. Figure 15-24	The origin of surface tension

Physlet® Illustrations

15-1 Archimedes' Principle
15-2 Blood Flow in Arteries

Suggested Readings

Denardo, B., Pringle, L., DeGrace, C. and McGuire, M., "When Do Bubbles Cause a Floating Body to Sink?" *American Journal of Physics* (October 2001), pp. 1064-1072. Describes qualitative lecture demonstrations that show that bubbles can sink a floating body, as well as a related quantitative experiment.

Eastlake, C., "An Aerodynamicist's View of Lift, Bernoulli, and Newton," *The Physics Teacher* (March 2002), pp. 166-173. Outlines alternative, but not conflicting, models to explain lift on an airfoil.

Gaffney, C., "The Hydrostatics of Trapped Bubbles in Fluids," *The Physics Teacher* (November 2000), pp. 458-460. A clever look at bubbles trapped in fluids and their effect on fluid level height.

Greene, N. and Dworsak, M., "Bernoulli at the Gas Pump," *The Physics Teacher* (September 2001), pp. 346-348. Discusses the basic physics of the shutoff sensing mechanism in a gas pump.

Noll, E., "Confronting the Buoyant Force," *The Physics Teacher* (January 2002), pp. 8-10. A description of a challenge problem involving the buoyant force, and student-designed methods for its solution.

Ouellette, J., "The Physics of . . . Foam," *Discover* (June 2002), pp. 26-27. A short article on the physics of soft matter.

Smith, D., "Flexural Strength in Windows During Hurricanes," *The Physics Teacher* (October 2000), pp. 400-402. Nice discussion of pressure changes and bulging windows during hurricanes.

Velasco, S., González, A., Román, F., and White, J., "A Simple Method for Measuring Atmospheric Pressure," *American Journal of Physics* (December 2002), pp. 1236–1237. Describes a simple, low-cost method for measuring atmospheric pressure.

Vorobieff, P. and Ecke, R., "Fluid Instabilities and Wakes in a Soap-Film Tunnel," *American Journal of Physics* (May 1999), pp. 394-399. Description of a two-dimensional hydrodynamic flow visualization system suitable for demonstrations and studies of a variety of fluid mechanics problems.

Waltham, C., "The Flight of a Balsa Glider," *American Journal of Physics* (July 1999), pp. 620-623. Analysis of the flight of a small toy glider, with comparison to full-sized aircraft.

Whitesides, G. and Stroock, A., "Flexible Methods for Microfluidics," *Physics Today* (June 2001), pp. 42-48. Applications for handling nanoliter quantities of fluids.

Notes and Ideas

Class time spent on material: Estimated:______ Actual:______________

Related laboratory activities:

Demonstration materials:

Notes for next time:

Chapter 16: Temperature and Heat

Outline

16-1 Temperature and the Zeroth Law of Thermodynamics
16-2 Temperature Scales
16-3 Thermal Expansion
16-4 Heat and Mechanical Work
16-5 Specific Heats
16-6 Conduction, Convection, and Radiation

Summary

Chapter 16 is the first of three chapters on thermal physics. The concepts of temperature, thermal energy, and heat are introduced. Applications of conservation of energy in this chapter include calorimetry as well as the connection between mechanical work and heat. Finally, processes of heat exchange are examined.

Major Concepts

By the end of the chapter, students should understand each of the following and be able to demonstrate their understanding in problem applications as well as in conceptual situations.

- Temperature
  - The zeroth law of thermodynamics
  - Temperature scales
  - Absolute zero
- Thermal expansion
  - Linear
  - Area
  - Volume
- Heat
  - Energy transfer
  - Mechanical work
  - Specific heats
  - Calorimetry
- Mechanisms of heat exchange
  - Conduction
  - Convection
  - Radiation

Teaching Suggestions and Demonstrations

This chapter lays the foundation for Chapters 17 and 18. Although the material can be covered fairly quickly, it is important to ensure that students understand the basics before moving on.

Sections 16-1 and 16-2

Students will most likely be familiar with all three **temperature scales** discussed in this chapter. Derive one of the conversion formulas with them to point out that both zero point and degree size have to be taken into account. The concept of **absolute zero** can be tricky. Figure 16-3 helps explain how the value of absolute zero is found. The **zeroth law of thermodynamics** is essentially common sense, but don't be tempted to skip over it. The concept is critical for an understanding of calorimetry, and students need to become comfortable with terminology such as "thermal contact" and "thermal equilibrium."

➲ DEMO **16-1** Make your own **thermometer**. Use a blank alcohol thermometer tube (or just put tape over the scale on a thermometer). (See Resource Information.) Put the tube in an ice-water bath and mark the level of the alcohol. Put the tube in a boiling-water bath and mark the level again. Define your own scale! Derive a conversion formula between your scale and the Celsius scale.

Section 16-3

Thermal expansion is the principle that causes a thermometer to work. As the liquid in the tube is heated, its volume expands, causing the liquid level to rise. Bridges and railroad tracks must be built with expansion joints to allow for varying temperatures, as shown in the photographs in Section 16-3. Figure 16-6 makes the very important point that when a container expands, the volume it encloses expands as well. Try Example 16-3 with your students to remind them to take into account both the expansion of the liquid and the expansion of the container holding it. As another example, ask students if they have ever held a jar with a tight lid under hot water in order to loosen the lid. Both the jar and the lid expand, but the lid expands more, making it easier to unscrew. Compare typical coefficients of thermal expansion for glass and metal in Table 16-1.

Sections 16-4 and 16-5

Conservation of energy is back again! **Heat** is energy transferred from one object to another. Mechanical energy can also be converted into heat. Remind students of the problems in Chapter 6 in which mechanical energy was not conserved in the presence of friction. Now you can explain that the energy wasn't really lost; it was just converted to thermal energy. Vigorously rubbing your hands together provides a quick demonstration of the conversion of mechanical to thermal energy. Discuss the units of heat, the **calorie** and the **kilocalorie**, and their relationship to the common energy unit, the joule.

Different substances require different amounts of heat to raise a unit mass through the same temperature difference. Water has a fairly high **specific heat**, and as a result, it takes larger additions or removals of thermal energy to change its temperature by much. Thus, oceans have moderating effects on the climate of coastal areas. **Calorimetry** is an application of the conservation of energy when substances with different temperatures are brought into thermal contact. If no heat escapes to the external environment, then the heat lost by the warmer substance(s) equals the heat gained by the cooler.

➲ DEMO **16-2** Heat metal shot to a known temperature and pour it into a **calorimeter** can filled with water at room temperature. (See Resource Information.) The shot releases heat and cools, while the can and water, which were initially in thermal equilibrium, warm up until the entire mixture reaches a stable temperature. To quantify the demonstration, measure the increase in temperature of the can and water and calculate the specific heat of the metal.

In the examples in this chapter, heat entering or leaving a substance results in a temperature change. In Chapter 17 students will see that phase changes also involve loss or gain of heat, but have no accompanying change in temperature.

Section 16-6

You may not have time to discuss the three mechanisms of **heat exchange** in detail, but it is still certainly worthwhile to introduce them conceptually. As examples, give several instances of heat exchange and have students name the mechanism(s) responsible in each case.

➲ DEMO 16-3 **Conduction, convection and radiation** can all be demonstrated using a pot filled with water on a hot plate. The bottom of the pot heats up due to conduction. Dye droplets squeezed into the water as it is heating show the convection currents. If you hold your hand close to the glowing hotplate, you can feel the radiation.

Resource Information

Transparencies

179. Figure 16-1	An illustration of the zeroth law of thermodynamics
180. Figure 16-2	A constant-volume gas thermometer
181. Figure 16-3	Determining absolute zero
182. Figure 16-4	Temperature scales
183. Figure 16-5	A bimetallic strip
184. Figure 16-6	Volume expansion
185. Figure 16-7	The unusual behavior of water near 4°C
186. Figure 16-8	The mechanical equivalent of heat
187. Figure 16-10	Countercurrent heat exchange in the human arm
188. Figure 16-11	Alternating land and sea breezes
189. Figure 16-12	The Thermos bottle

Physlet® Illustrations

16-1 Thermal Expansion
16-2 Specific Heat
16-3 Thermal Conductivity

Suggested Readings

Denardo, B., "Temperature of a Lightbulb Filament," *The Physics Teacher* (February 2002), pp. 101-105. Investigates the error introduced by the assumption that the variation of resistivity with temperature is linear.

Gash, P., "So You Thought a Glass Thermometer Measured Temperature," *The Physics Teacher* (February 2002), pp. 74-76. An application of calorimetry, which includes the heat absorbed or lost by the glass of the thermometer.

Lee, W., Gilley, H. and Caris, J., "Finding the Surface Temperature of the Sun Using a Parked Car," *American Journal of Physics* (November 1997), pp. 1105-1109. An activity involving both quantitative observations and theoretical analysis for estimating the surface temperature of the Sun.

McGee, H., McInerney, J. and Harrus, A., "The Virtual Cook: Modeling Heat Transfer in the Kitchen," *Physics Today* (November 1999), pp. 30-36. An interesting though advanced article on what happens in the cooking process.

Reif, F., "Thermal Physics in The Introductory Physics Course: Why and How to Teach It From a Unified Atomic Perspective," *American Journal of Physics* (December 1999), pp. 1051-1062. An argument for and description of using an atomic approach for teaching thermal physics.

Silverman, M. and Silverman, C., "Cool in the Kitchen: Radiation, Conduction, and the Newton 'Hot Block' Experiment," *The Physics Teacher* (February 2000), pp. 82-88. Valuable information about heat loss and thermal properties.

Wolbarst, A., "Effective Thermal Conduction Model for Estimating Global Warming," *American Journal of Physics* (October 1999), pp. 885-890. Illustrates thermal conduction and radiation using the important example of greenhouse warming of the Earth.

Materials

Thermometers are available from a variety of sources, including Fisher Scientific, model number S41577A or S41580 (ungraduated).

A student calorimeter is available from Fisher Scientific, model number S41746.

Notes and Ideas

*Class time spent on material: Estimated:*_______ *Actual:*__________

Related laboratory activities:

Demonstration materials:

Notes for next time:

Chapter 17: Phases and Phase Changes

Outline

Summary

Chapter 17 continues the study of thermodynamics begun in Chapter 16. Ideal gases are defined and used to present the kinetic theory of gases, which connects macroscopic variables (temperature and pressure) to microscopic variables (molecular speed and mass). Mechanical properties of solids are discussed. The last three sections of this chapter deal with phase changes and the relationship between heat given off or absorbed by a material and the corresponding temperature or phase change.

Major Concepts

By the end of the chapter, students should understand each of the following and be able to demonstrate their understanding in problem applications as well as in conceptual situations.

- Ideal gases
  - Definition
  - Equation of state
- Mole (Avogadro's number)
- Kinetic theory of gases
  - Molecular speed distribution
  - Kinetic energy, pressure, temperature
  - Internal energy of a gas
- Mechanical properties of solids
  - Linear expansion
  - Shear deformation
  - Bulk modulus
  - Stress and strain
- Phase equilibrium
- Latent heat
- Phase changes and energy conservation

Teaching Suggestions and Demonstrations

Students who have had chemistry will find some familiar material in this chapter. It is useful to find out how many students have seen this material before. You may be able to skim through the section on ideal gases and latent heat and give more time to the less familiar parts of the chapter.

Sections 17-1 and 17-2

Plan to spend a little time on the definition of an **ideal gas** even if all the students have seen the equation of state before. Point out how an ideal gas differs from a real gas, and why the concept of an ideal gas is useful. Go over the **constant-volume gas thermometer** (Figure 17-1) and connect it to the concepts from Chapter 16.

➲ **DEMO 17-1** The form of the **equation of state for an ideal gas** is deduced quite nicely from observations, as described in Section 17-1. Bring in a basketball and an air pump and go through the two exercises discussed.

Even if most of the students have had chemistry, you will need to discuss the concepts of the **mole** and **Avogadro's number**. Be sure to go over the relationship between the **Boltzmann constant (k)** and the **universal gas constant (R)**. Active Example 17-1 will help students develop an intuitive feel for the "size" of a mole of air.

If you are pressed for time, you can skip the individual discussions of **Boyle's law** and **Charles's law**. However, you do need to go over the definitions of **isotherms** and **isobars**, since they will be needed again in Chapter 18.

The **kinetic theory of gases** connects the macroscopic quantities that describe the gas as a whole to the microscopic quantities that describe the individual molecules of the gas. Students sometimes have trouble believing that this can actually be done; be prepared to present evidence and argue the case. Go over Conceptual Checkpoint 17-2 to help students relate pressure, number of molecules and rms speed.

➲ **DEMO 17-2** You can use the evidence of the scarcity of low atomic weight gases (hydrogen and helium) in the atmosphere as an argument in support of the **kinetic theory of gases**. Since the masses of these gas molecules are small compared to the molecular masses of nitrogen or oxygen, the rms velocity of hydrogen or helium molecules will be larger than that of other molecules at the same pressure. In fact, the rms velocity of hydrogen molecules at atmospheric pressure and normal temperature is on the same order of magnitude as escape velocity! (See Figure 17-20 and the Real World Physics discussion on planetary atmospheres.) Do the calculation, and discuss its implications with the class.

The **internal energy of an ideal gas** is also mentioned in this section. Be sure to cover the definition here, as it will come up again in Chapter 18.

Section 17-3

This section contains some really fascinating material, but if time is limited, you can skip it entirely. If you do have time to cover it, **stretching and Young's modulus**, and **shear deformation and the shear modulus** are explained simply and quickly in the text. The discussion of the change in the volume of a solid with increasing pressure includes the definition of the **bulk modulus**. Do Active Example 17-4 and use it to point out the very different behavior of gases, liquids, and solids when external pressures are applied. **Stress, strain**, and **elastic deformation** are also presented clearly; go over Figure 17-13 and talk about the **elastic limit**. Make sure students understand that strain is a ratio and therefore has no units.

➲ **DEMO 17-3** Use a foam cube to demonstrate the three types of **stress**.

Sections 17-4 through 17-6

Much of Section 17-4 can also be skipped with no loss of continuity, although you should try to at least introduce phase diagrams (see Figure 17-16.) The discussion on **evaporative cooling** is interesting and informative. Plan to spend more time on Sections 17-5 and 17-6. Students usually have trouble with the concept that during a (quasi-static) **phase change** a system does not change temperature. Go over the three latent heats: **latent heat of fusion**, **latent heat of vaporization**, and **latent heat of sublimation**, and the phase changes to which they apply. Discuss Figure 17-21 in detail and go over Conceptual Checkpoint 17-5.

➲ **DEMO 17-4** For this demo you will need a saucepan or Pyrex beaker, crushed ice (as cold as you can get it), a hot plate, and a thermometer. Fill the saucepan or Pyrex beaker with the ice, insert the thermometer and have a student record the temperature. Place the pan or beaker on the hot plate. Turn the burner on low, and have a student check the temperature of the ice (which will eventually melt) every two minutes (or one minute, or five minutes – as long as the time interval is constant). Once it heats up, the hot plate will deliver heat to the pan at (approximately!) a constant rate. If you plot **temperature vs. time**, you should be able to reproduce Figure 17-21. This is a little crude, but it will give students a concrete idea of what Figure 17-21 represents. You can take the temperature of the water all the way up to the boiling point.

➲ **DEMO 17-5** Make ice cream in an electric freezer. Use the Real World Physics discussion at the end of Section 17-5 to connect this exercise to energy transfer and latent heat. Point out that the cranking, whether done by machine or by human power, serves to break up the ice crystals and keep them from getting too large as they freeze. This is a good exercise to do on the last day of class – you can set it up at the beginning of class, and it should be ready to eat by the end!

Section 17-6 connects the concept of specific heat from Chapter 16 with the latent heats from Chapter 17 to give a general **conservation of energy** equation for isolated systems. You can treat this concept as $\Delta Q = 0$, or as $Q_{lost} = Q_{gained}$. Either way, you will need to work many problems such as Example 17-6 that involve both temperature and phase changes.

Resource Information

Transparencies

190. Figure 17-1	A constant-volume gas thermometer
191. Figure 17-5	Ideal-gas isotherms
192. Figure 17-6	Volume versus temperature for an ideal gas at constant pressure
193. Figure 17-7	Force exerted by a molecule on the wall of a container
194. Figure 17-8	The Maxwell speed distribution
195. Figure 17-9	Most probable, average, and rms speeds
196. Figure 17-10	Stretching a rod
197. Figure 17-11	Shear deformation
198. Figure 17-13	Stress versus strain
199. Figure 17-14	A liquid in equilibrium with its vapor
200. Figure 17-15	The vapor-pressure curve for water
201. Figure 17-16	A typical phase diagram

202. Figure 17-21 Temperature versus heat added or removed

Physlet® Illustrations

17-1 Ideal Gas Law
17-2 Kinetic Theory
17-3 Young's Modulus and Tensile Strength

Suggested Readings

Capitolo, M., "Phase-Change Demonstration - Instant Gratification," *The Physics Teacher* (September 1998), p. 349. A simple but dramatic phase-change demonstration.

Graham, M., "Investigating Gases' Masses in Impecunious Classes," *The Physics Teacher* (March 2002), pp. 144-147. Directions for building and using an inexpensive apparatus to measure molecular mass of a gas to an accuracy of about 10 percent.

Gutiérrez, G. and Yáñez, J., "Can an Ideal Gas Feel the Shape of its Container?," *The American Journal of Physics* (August 1997), pp. 739-743. Examination of thermodynamic quantities of an ideal gas enclosed in a finite container.

Hart, F., "A Microcomputer-Based Phase-Change Experiment," *The Physics Teacher* (February 1998), pp. 98-99. A laboratory exercise for examining a phase change.

Schmidhuber, C., "On Water, Steam, and String Theory," *The American Journal of Physics* (November 1997), pp. 1042-1050. A detailed description of a second-order phase transition between water and steam, with applications in particle physics and string theory.

Thompson, D., "Derivation of the Ideal Gas Law from Kinetic Theory," *The Physics Teacher* (April 1997), pp. 238-239. A short derivation of the ideal gas law using collisions of particles with container walls.

Velasco, S., Faro, J. and Roman, F., "An Experiment for Measuring the Low Temperature Vapor Line of Water," *American Journal of Physics* (December 2000), pp. 1154-1157. A description of a method to measure the vapor pressure curve of water below the normal boiling temperature.

Notes and Ideas

*Class time spent on material: Estimated:*______ *Actual:*____________

Related laboratory activities:

Demonstration materials:

Notes for next time:

Chapter 18: The Laws of Thermodynamics

Outline

18-1 The Zeroth Law of Thermodynamics
18-2 The First Law of Thermodynamics
18-3 Thermal Processes
18-4 Specific Heats for an Ideal Gas: Constant Pressure, Constant Volume
18-5 The Second Law of Thermodynamics
18-6 Heat Engines and the Carnot Cycle
18-7 Refrigerators, Air Conditioners, and Heat Pumps
18-8 Entropy
18-9 Order, Disorder, and Entropy
18-10 The Third Law of Thermodynamics

Summary

Chapter 18 presents the fundamental laws of thermodynamics. The zeroth law, first introduced in Chapter 16, defines temperature. The first law applies conservation of energy to thermodynamic systems. The second law defines the concept of entropy, and the third law gives the limiting value of temperature, absolute zero.

Major Concepts

By the end of the chapter, students should understand each of the following and be able to demonstrate their understanding in problem applications as well as in conceptual situations.

- The zeroth law of thermodynamics
- The first law of thermodynamics
  - Definitions of Q, W and ΔU
  - Sign conventions
  - Specific heat at constant pressure
  - Specific heat at constant volume
- The second law of thermodynamics
  - Heat engines
  - Carnot's theorem
  - Refrigerators and heat pumps
- Entropy
- The third law of thermodynamics

Teaching Suggestions and Demonstrations

Thermodynamics remains a great mystery to many students in introductory physics. Parts of it (such as the zeroth law) seem obvious, but other parts (such as entropy) seem very difficult. Use lots of examples and try to tie the concepts to practical applications.

Sections 18-1 through 18-3

Go over the **zeroth law** again for completeness. (It was covered in Chapter 16.) Then start on the **first law of thermodynamics**. Figures 18-1 and 18-2 are wonderful illustrations of the special cases that help deduce the first law. Be sure to go over the definitions and sign conventions for $\boldsymbol{Q}$**,** $\boldsymbol{W}$**,** and $\boldsymbol{\Delta U}$. It is important for students to understand the differences between Q, which represents the **flow of thermal energy**; W, which represents **physical work involving a force**; and U, which is a **state function**, determined by the state variables of pressure, temperature, and volume.

The idea of a **quasi-static or reversible process** will be new to the students. Discuss the definitions of reversible and **irreversible processes** thoroughly.

As you present **constant pressure**, **constant volume**, and **constant temperature processes**, use the illustrations in Figures 18-4 and 18-6 and in Example 18-3. Figures 18-5, 18-7, 18-8, and 18-9 introduce ***PV*** **plots**, which can illustrate many thermodynamic processes. Make sure students are comfortable interpreting and sketching these plots. Work many examples, including Examples 18-2 and 18-3 and Active Example 18-1.

In an **adiabatic process** no heat is exchanged with the environment, but the temperature of the system can change. This concept is difficult for students, who often equate heat exchange with a change in temperature. Use Figure 18-10 to discuss adiabatic processes and go over Conceptual Checkpoint 18-2 to check understanding. The Real World Physics discussion connecting **adiabatic heating and diesel engines** presented in Section 18-3 is a good discussion starter. Table 18-2 gives a compact summary of the definitions presented in these sections. As a review, have students apply the first law to each different type of process, and see which term, if any, is zero.

Section 18-4

This section can be skipped without loss of continuity. If you choose to cover it, you can discuss the differences between **specific heat at constant pressure** and **specific heat at constant volume**, based on what the students now know about thermodynamic processes. Figures 18-12 and 18-13 are very helpful. The treatment of adiabatic processes using C_v and C_p is interesting, but the math involved (a variable exponent) may be beyond some of the students. The discussion of **adiabatic heating and cooling** in the Real World Physics example of **rain shadows** is a nice application of thermodynamic principles. Plan to take a short time to go over it in class.

Sections 18-5 through 18-7

The **second law of thermodynamics** gives a direction to the flow of energy in a thermodynamic process.

➲ **DEMO 18-1** For a very simple demonstration, bring several (around 10) beakers to class. Fill each partially with water and put some ice in each one. Ask the class to predict the number of beakers in which the ice will melt and the number of beakers in which the water will freeze. Most of the class will realize that, of course, the ice will melt in all of the beakers. You can use this as a discussion starter on the **second law**.

A **heat engine** converts energy into work, with (ALWAYS) waste heat ejected to the environment. Go over the steam engine (see Figure 18-15) and the schematic diagram of a heat engine in Figure 18-16. The concepts of a **hot reservoir** and a **cold reservoir** may be difficult for some students. Define the **efficiency** of a heat engine and point out that you can never have a heat engine that is 100 percent efficient. Work Example 18-6 and several other examples from the homework problems.

Carnot's theorem gives the upper limit on the efficiency of a heat pump operating between two constant-temperature reservoirs. If the students have understood the difference between reversible and irreversible processes, they should follow the theorem without too much trouble. Be sure to explain the equation for the **maximum efficiency of a heat engine** and work several examples.

Refrigerators, **air conditioners** and **heat pumps** all use work to move energy from a cold reservoir to a hot reservoir. Use Figure 18-17 to illustrate the similarities and differences between these devices and the heat engine. Define the **coefficient of performance for a refrigerator** and the **coefficient of performance for a heat pump**. Conceptual Checkpoint 18-5 has a good discussion on how an air conditioner works.

Sections 18-8 and 18-9

Entropy is a difficult concept for students to grasp. They can learn to do calculations, and they can develop intuition about which processes create more **disorder**. Try to be as concrete as possible when describing entropy and its applications in physics. If you are short on time, you can just discuss the concept of entropy and skip the equations.

➲ **DEMO 18-2** Bring in a new deck of cards. Show the class that the cards are sorted in order by suit. The new deck is an **ordered system**. Shuffle the cards, and in the process spray them all over the room. (This usually takes the students by surprise.) Ask if the new configuration of cards (all over the room) has more or less order than the original configuration. Can the original configuration be restored? Yes, but (relatively easy) work must be done!

➲ **DEMO 18-3** Bring in a clear beaker of water and a bottle of food coloring. Place the beaker where it can be seen by the whole class and where it can stay for the entire class period. Point out that the system of the beaker of water and the bottle of food coloring is an ordered system. Place one or two drops of food coloring in the beaker, being careful not to bump the beaker or stir the mixture. The food coloring will disperse throughout the beaker by the end of class. Is the mixture more or less disordered than the original system? Can the original configuration be restored? No, not without using a tremendous amount of energy!

Discuss the definition of ΔS and point out that it is the **change in entropy** in a process that is important, not the total entropy of a system. Work Example 18-9 and Active Example 18-3 in class and go over Conceptual Checkpoint 18-6. The Real World Physics discussion of **entropy and living systems** is well worth some class time.

Section 18-10

The **third law of thermodynamics** uses thermal contact and the transfer of energy to argue that absolute zero is a limiting temperature. Objects can get close to absolute zero, but never all the way there in a finite number of steps.

Resource Information

Transparencies

203.	Figure 18-1	The internal energy of a system
	Figure 18-2	Work and internal energy
204.	Figure 18-3	An idealized reversible process
205.	Figure 18-4	Work done by an expanding gas
	Figure 18-5	A constant-pressure process
206.	Figure 18-6	Adding heat to a system of constant volume
	Figure 18-7	A constant-volume process
207.	Figure 18-8	Isotherms on a *PV* plot
208.	Figure 18-9	An isothermal expansion
209.	Example 18-3	Heat Flow
210.	Figure 18-10	An adiabatic process
211.	Conceptual Checkpoint 18-2	
212.	Table 18-2	Thermodynamic Processes and Their Characteristics
213.	Figure 18-14	A comparison between isotherms and adiabats
214.	Figure 18-15	A schematic steam engine
215.	Figure 18-17	Cooling and heating devices

Physlet® Illustrations

18-1 Work Done at Constant Temperature
18-2 Work Done at Constant Pressure
18-3 Carnot Cycle

Suggested Readings

Graham, D. and Schacht, D., "Simple Estimate of the Human Metabolic Rate," *American Journal of Physics* (June 2001), pp. 723-724. An experiment involving energy and entropy that will be of particular interest to students in the pre-health professions.

Leff, H., "What If Entropy Were Dimensionless?," *American Journal of Physics* (December 1999), pp.1114-1122. Conclusions from an examination of the puzzling aspects of the dimensions of entropy.

Lieb, E. and Yngvason, J., "A Fresh Look at Entropy and the Second Law of Thermodynamics," *Physics Today* (April 2000), pp. 32-37. An accessible and different approach to understanding entropy without applying statistical mechanics.

Loverude, M., Kautz, C. and Heron P., "Student Understanding of the First Law of Thermodynamics," *American Journal of Physics* (February 2002), pp.137-148. Report on an investigation which indicated that students often fail to differentiate among the concepts of heat, temperature, work and internal energy. Includes implications for instruction.

Saslow, W., "An Economic Analogy To Thermodynamics," *American Journal of Physics* (December 1999), pp. 1239-1247. An interesting analogy, especially if there are students from a variety of majors in your class.

Schoepf, D., "A Statistical Development of Entropy for the Introductory Physics Course," *American Journal of Physics* (February 2002), pp.128-136. A model for introducing the statistical basis for the definition of entropy, including a correspondence between the statistical and the classical Clausius definitions.

Styer, D., "Insight into Entropy," *American Journal of Physics* (December 2000), pp. 1090-1096. Several examples from statistical mechanics demonstrating the qualitative character of entropy.

Velasco, S., Román, F. and Faro, J., "A Simple Experiment for Measuring the Adiabatic Coefficient of Air," *American Journal of Physics* (July 1998), pp. 642-645. Instructions for performing an experiment to measure compressibility and the adiabatic coefficient of air.

Yeo, S. and Zadnik, M., "Introductory Thermal Concept Evaluation: Assessing Students' Understanding," *The Physics Teacher* (November 2001), pp. 496-504. Presentation of a multiple-choice evaluation instrument designed to test student understanding of thermal concepts.

Notes and Ideas

*Class time spent on material: Estimated:*______ *Actual:*______________

Related laboratory activities:

Demonstration materials:

Notes for next time:

Chapter 19: Electric Charges, Forces, and Fields

Outline

Summary

The electric force affects our daily lives more than any other force. Chapter 19 introduces the concepts of charge, Coulomb's law, and the electric field. The electric properties of various materials are discussed. Charging by induction, electric flux, and Gauss's law are also presented.

Major Concepts

By the end of the chapter, students should understand each of the following and be able to demonstrate their understanding in problem applications as well as in conceptual situations.

- Electric charge
  - Positive and negative
  - Quantization of charge
  - Conservation of charge
- Electric properties of materials—insulators and conductors
- Coulomb's law
- The electric field
  - Definition
  - Field lines
- Induction
- Gauss's law
  - Electric flux

Teaching Suggestions and Demonstrations

Electric charges and the basic rules that govern their behavior (opposite charges attract; like charges repel) are familiar to most students. The challenge for this chapter is to build on that intuition so more complex situations can be treated.

Section 19-1

Emphasize to the students that the labels of **positive and negative charge** are arbitrary (assigned by Benjamin Franklin!) and that they work well for most situations. Charge is **quantized**, occurring in "bits" of e, the magnitude of the fundamental charge on the electron or proton.

➲ **DEMO 19-1** Rub a glass or Lucite rod with animal fur (or wool or silk). (See Resource Information.) You will charge the rod. Use the charged rod to pick up small pieces of paper. If it is winter, you can also do this demo with a plastic comb and clean hair, or a balloon and clean hair. (See Table 19-1 for a list of materials that can be charged by rubbing.) If you are teaching this class in the summer you are probably out of luck – very few electrostatic demonstrations "work" during the summer due to higher levels of humidity!

➲ **DEMO 19-2** Do the demonstration involving a charged balloon and a stream of water shown in the photograph in Section 19-1. The movement of the stream of water is very dramatic!

Be sure to go over the **conservation of charge**. Remind the students that when we charge objects as in the above demonstration, we are transferring charge from one object to another, not creating charge. Discuss how **polarization of charge** enables these demonstrations to work. (See Figure 19-5.)

Section 19-2

Go over the definitions of **insulators** and **conductors**. Students should understand that in insulators any excess charge is not free to move, while in conductors any excess charge is free to move about the surface. **Semiconductors** are intermediate between conductors and insulators; their properties can be controlled by adding small amounts of other materials. There is a nice discussion at the end of this section of **photoconductive materials** and their application to laser printers and photocopiers that is worth class time.

Section 19-3

Present the definition of **Coulomb's law**, which is true for point charges and spherically symmetric charge distributions. Be sure to point out the similarities between Coulomb's law and the law of universal gravitation. Note that the gravitational force is always attractive, but the **electrostatic force** can be attractive or repulsive. (See Figure 19-7.) Conceptual Checkpoint 19-2 should generate some interesting class discussion about force and acceleration. The calculation of the **relative size of the electrostatic and gravitational forces** in this section will be startling to the students and is good to go through in class. Example 19-1 ties the electrostatic force to the centripetal force.

In general, students will need to be reminded about **vectors, vector addition**, and the **superposition of forces**. Plan to work many examples of force calculations. Conceptual Checkpoint 19-3 is a good test of student understanding of the vector addition of forces.

Sections 19-4 through 19-6

The idea of a **force field** will be alien to most students. Be as concrete as possible in describing this concept. Figure 19-9 will help. Go over the definition of the field (force per unit charge) and compare it to the definition of g (force per unit mass).

➲ **DEMO 19-3** On the board or on an overhead, use the idea of a small **positive test charge** to map out the **force field** from a positive charge q located at the origin. Use arrows to indicate the direction and magnitude of the field at many points. (Figure 19-11 will give you a start.) Remind the students that the field is three-dimensional, not flat like the board. Once you have the field set up, be sure to discuss the direction of the force on a **negative test charge**.

Note that the **superposition of electric fields** from several charges involves vector addition. (See Figure 19-12.) Plan to spend some time on this concept in class and work several examples. Go over Conceptual Checkpoint 19-4 to assess student understanding.

The above demonstration leads right into a discussion of **electric field lines**. The rules for drawing electric field lines are given at the beginning of Section 19-5. Figures 19-14 through 19-16 illustrate several of the key concepts. Figure 19-17 is a picture of the ideal **parallel-plate capacitor**, which will be discussed in Chapters 20, 21 and 24 in more detail. Example 19-6 is a good exercise to work in class and the following discussion about televisions and ink-jet printers will be of interest to the students.

Section 19-6 begins with an excellent description of how **excess charge** is distributed on a conductor. This leads logically into a discussion of electrostatic **shielding** and **induced charge**. You will need to spend some time on both of these concepts, as students find them somewhat confusing. Go over the Real World Physics example on **shielding** and Figure 19-21. Use Figures 19-19 and 19-20 to talk about **electric fields at conductor surfaces**. Be sure to point out how they differ from electric fields at insulator surfaces, which are not required to be perpendicular to the surface. Take the students step-by-step through the **charging by induction process**; the logical progression of steps will make sense to most students.

Section 19-7

Electric flux, like the electric field, is a difficult concept for many students because they can't see it. The more visual you can make your presentation of this section, the better.

➲ **DEMO 19-4** Bend some wire (coat hangers do nicely) into a square loop. Draw parallel electric field lines on the board. Hold up the loop so that its face is perpendicular to the field lines. Discuss the **normal to the surface**, and point out that in this position the normal is parallel to the field lines and the most field lines pass through the loop. Rotate the loop until the normal is perpendicular to the field lines. Point out that in this position no field lines pass through the loop. Once you have discussed how the electric field lines pass through (or just pass by) the loop, you can define the electric flux. Be sure to talk about **electric flux through closed surfaces** and give several examples.

You can present **Gauss's law** by talking about "sources" (faucets) and "sinks" (drains) of water. If you place a closed surface made of screen wire in a flowing stream, the amount of water entering the surface is equal to the amount of water leaving the surface. The "net flux" of water through the surface is zero. If you place the closed surface over the end of a faucet, there is a **net positive flux** through the surface. If you place the closed surface over a drain, there is a **net negative flux** through the surface. Gauss's law works just this way for electric fields. Positive charges are "sources" of electric field lines; negative charges are "sinks" of electric field lines. If a closed surface (usually imaginary) encloses no charge, then the net flux of electric field lines through the surface is zero. If it contains a net positive or negative charge, then the net flux of electric field lines is positive or negative, respectively. Go over Conceptual Checkpoint 19-6 and the two primary types of Gaussian surfaces (spheres and cylinders) and work several examples. (See Figures 19-25 and 19-26.)

Resource Information

Transparencies

216. Figure 19-5	Electrical polarization
217. Figure 19-7	Forces between point charges

218. Figure 19-8 Superposition of forces
219. Figure 19-9 An electrostatic force field
220. Figure 19-12 Superposition of the electric field
221. Figure 19-14 Electric field lines for a point charge
222. Figure 19-15 Electric field lines for systems of charges
223. Figure 19-16 The electric field of a charged plate
224. Figure 19-17 A parallel plate capacitor
225. Figure 19-18 Charge distribution on a conducting sphere
226. Figure 19-19 Electric field near a conducting surface
227. Figure 19-20 Intense electric field near a sharp point
228. Figure 19-21 Shielding works in only one direction
229. Figure 19-22 Charging by Induction
230. Figure 19-23 Electric flux
231. Figure 19-24 Electric flux for a point charge
232. Figure 19-25 Gauss's law applied to a spherical shell
233. Figure 19-26 Gauss's law applied to a sheet of charge
234. Active Example 19-3 Find the Electric Field

Physlet® Illustrations

19-1 Coulomb's Law
19-2 Electric Fields
19-3 Electric Field in a Parallel-Plate Capacitor

Suggested Readings

Avants, B., Soodak, D. and Ruppeiner, G., "Measuring the Electrical Conductivity of the Earth," *American Journal of Physics* (July 1999), pp. 593-598. An experiment for measuring the electrical conductivity of the Earth to yield information about what is underground.

Bracikowski, C., "Graphical Analysis of Electric Fields of Dipoles and Bipoles," *The Physics Teacher* (January 2000), pp. 20-21. A qualitative explanation of the dependence of field strength on distance.

Cortel, A., "Demonstrations of Coulomb's Law with an Electronic Balance," *The Physics Teacher* (October 1999), pp. 447-448. A method of measuring weak repulsive forces between electric charges.

Gelbart, W., Bruinsma, R., Pincus, P. and Parsegian, V., "DNA-Inspired Electrostatics," *Physics Today* (September 2000), pp. 38-44. Research concerning electrostatic phenomena and electrostatic interactions between biomolecules that will be of particular interest to biology majors and students in the pre-health professions.

Harrington, R., "Getting a Charge Out of Transparent Tape," *The Physics Teacher* (January 2000), pp. 23-25. Two experiments using charging by adhesion-separation.

Jefimenko, O., "Dynamic Electric Field Maps of Point Charge Moving with Constant Velocity," *The Physics Teacher* (March 2000), pp. 154-157. An alternative way to graphically represent the electric fields of moving charges.

Materials

One electrostatics demonstration kit is available from Fisher Science Education, model number S57200-1. Other electrostatic demonstration equipment is available from Fisher and other science equipment companies.

Notes and Ideas

Class time spent on material: Estimated:______ Actual:______________

Related laboratory activities:

Demonstration materials:

Notes for next time:

Chapter 20: Electric Potential and Electric Potential Energy

Outline

Summary

Chapter 20 continues the study of electricity begun in Chapter 19. Electric potential energy and electric potential are defined. Conservation of energy is applied to electrical systems. Equipotential surfaces and their relationships to the electric field and electric field lines are covered. The chapter ends with a discussion of capacitors with and without dielectrics and how they store electrical energy.

Major Concepts

By the end of the chapter, students should understand each of the following and be able to demonstrate their understanding in problem applications as well as in conceptual situations.

- Electric potential energy
- Electric potential
  - Definition
  - Equipotential surfaces
- Capacitors
  - Definition
  - Dielectrics
  - Electrical energy storage

Teaching Suggestions and Demonstrations

Students generally understand the concept of electrical potential energy. Electrical potential is more difficult for them. You may need to review the definition of work from Chapter 7. Capacitors also seem to give students trouble. Work as many examples as possible.

Sections 20-1 through 20-3

Use the correspondence between the electric and gravitational forces (both forces are conservative, and both have the same inverse-square dependence) to argue for an **electric potential energy**. Be very careful to remind students (again!) that while the gravitational force is always attractive, the electric force can be attractive or repulsive. This affects the potential energy. Go over the calculation of the work done on a test charge (see the beginning of Section 20-1). Point out that (just like in the gravitational case) it is the **change in potential energy** that is significant, not the absolute value.

Go over the definition of the **electric potential**, the change in potential energy per unit charge, and its units. Plan to work several examples before moving on to the connection between the **change in the**

electric potential with position and the **electric field**. Students who have had some calculus will catch on to this concept relatively quickly; it will take more time for the students who have had no calculus. Work Example 20-1 and Active Example 20-1 and go over Conceptual Checkpoint 20-1. Work several other examples in class to give the students practice with this concept.

Students should remember the **work-energy theorem** from Chapter 7 and the **conservation of mechanical energy** from Chapter 8. (You may need to review these briefly.) The application of these ideas to the electric force and the electric potential energy is straightforward. Again, remind students that they must be very careful with signs, since the electric force can be attractive or repulsive. Active Example 20-2 and the paragraph following it provide good structure for a class discussion. Use Conceptual Checkpoint 20-2 to check for understanding.

The **electric potential of a point charge** is more difficult to describe without calculus because the field decreases as $1/r^2$. The qualitative argument at the beginning of Section 20-3 is useful. Be sure to point out to the students that ***V = 0* for a point charge at *r = infinity*** is a *choice*, and that (again) it is the change in the potential that is significant. Also note that the sign of the potential depends on the sign of the charge. (See Figure 20-5.)

If more than one charge is present, the net electric field is the *vector* sum of the electric fields from all the charges. Electric potential is a scalar; if more than one charge is present, the potential is the *algebraic* sum of the potential created by each charge. This is the **principle of superposition for potentials**. The same principle works for the electric potential energy, since energy is also a scalar. Go over Conceptual Checkpoint 20-3 and Example 20-4 in class.

Section 20-4

The analogy of the contour map at the beginning of this section works well in the introduction of **equipotential surfaces.**

➲ **DEMO 20-1** If possible, obtain a contour map of your campus (or your area). Make a transparency of part of it, and go over it with the class. Point out (as in the first paragraph of Section 20-4) that the curves represent different altitudes and that there is a constant altitude difference between contour lines. Closely spaced curves represent hills, or places where the altitude changes rapidly. Widely spaced curves represent more gentle changes in altitude.

Also go over transparencies of Figures 20-5 and 20-6 in class. Point out that Figure 20-6 is a two-dimensional representation of Figure 20-5, and that the circles represent **lines of constant potential** (in three dimensions these would be **surfaces of constant potential**). There is a constant potential difference between contour lines, just as in the contour map above. In the region where the circles are close together, the potential is changing rapidly. In the region where the circles are farther apart, the potential is not changing as fast.

Be sure to tell the students that the **electric field points in the direction of decreasing electric potential,** and that the **electric field lines are perpendicular to the equipotential surfaces.** Remind them that when the electric field lines are close together, the electric field is large. Go over Conceptual Checkpoint 20-4 and Figures 20-7 and 20-8 for good applications of these ideas.

The concept that the **surface of a conductor is an equipotential surface** is relatively easy for beginning physics students to understand. The implications of that concept for charge density of surfaces with different radii of curvature are much more difficult to grasp. Figures 20-9 and 20-10 are graphical representations of the **concentration of charge on surfaces with smaller radii**. Discuss them thoroughly and apply this concept to the development of lightning rods.

Sections 20-5 and 20-6

The definition of **capacitance** is deceptively simple. Many students have a difficult time with this concept. Go over the definition of a **capacitor**, then the definition of capacitance, pointing out that the capacitance serves as the constant of proportionality between the voltage applied to a capacitor and the amount of charge it can hold. Go through the derivation of the capacitance of a **parallel-plate capacitor**. Discuss how the capacitance depends on the area of the capacitor and the distance between the plates. Conceptual Checkpoint 20-5 is good for review.

It helps to remind students that **dielectrics** are just insulators placed between the plates of a capacitor. Figure 20-15 is a great diagram of what happens in the dielectric as it is moved between the capacitor plates. Since the field inside the dielectric is less than the field outside (due to the polarization of charge in the dielectric), it makes sense to divide the outside field strength by the **dielectric constant** to get the inside field strength. If you can convince the students that this is true, the rest of the derivation follows. Go through Example 20-6 in class. The Real World Physics application on computer keyboards is quite interesting, as is the application on the Theremin. If your students have heard the "spooky" music played for early science-fiction movies, they have heard a Theremin. (See Resource Information.)

➲ **DEMO 20-2** Bring in several capacitors of various sizes and shapes. Discuss why they are made the way they are. Point out that "normal" size capacitors have capacitances measured in the picofarad or microfarad range.

Because capacitors store charge, they also store electrical energy. Discuss the derivation of the **electrical energy stored in a parallel-plate capacitor** in Section 20-6. The Real World Physics applications on the electronic flash and the defibrillator are both good. Work Example 20-7 in class. Be sure to cover the hazards of capacitors and review the idea that the energy stored in the capacitor is stored in the electric field between the plates.

Resource Information

Transparencies

235. Figure 20-1	Change in electric potential energy
236. Figure 20-3	The electric potential for a constant electric field
237. Figure 20-5	The electric potential of a point charge
238. Figure 20-6	Equipotentials for a point charge
239. Figure 20-7	Equipotential surfaces for a uniform electric field
240. Figure 20-8	Equipotential surfaces for two point charges
241. Figure 20-9	Electric charges on the surface of ideal conductors
242. Figure 20-10	Charge concentration near points
243. Figure 20-11	The electrocardiograph
244. Figure 20-15	The effect of a dielectric on the electric field of a capacitor

Physlet® Illustrations

20-1 Work and Potential Difference
20-2 Equipotential Surfaces

Suggested Readings

Beichner, R., "Visualizing Potential Surfaces with a Spreadsheet," *The Physics Teacher* (February 1997), pp. 95-97. Presents a technique for developing three-dimensional plots of potential surfaces.

Moelter, M., Evans, J. and Elliott, G., "Electric Potential in the Classical Hall Effect: An Unusual Boundary-Value Problem," *American Journal of Physics* (August 1998), pp. 668-677. Detailed presentation of a solution to a nonstandard boundary value problem in a Hall effect experiment.

Parker, G., "Electric Field Outside a Parallel Plate Capacitor," *American Journal of Physics* (May 2002), pp.502-507. An interesting topic to mention, although it is beyond the scope of most introductory-level physics courses.

Saslow, W., "Voltaic Cells for Physicists: Two Surface Pumps and an Internal Resistance," *American Journal of Physics* (July 1999), pp. 574-583. A discussion of the basic properties of voltaic cells considering the chemical reactions that take place at the electrode–electrolyte interfaces.

Skeldon, K., Reid, L., McInally, V., Dougan, B. and Fulton, C., "Physics of the Theremin," *American Journal of Physics* (November 1998), pp. 945-955. A discussion of some of the interesting physics behind the design and operation of the Theremin electronic musical instrument.

Zimmerman, N., "A Primer on Electrical Units in the Système International," *American Journal of Physics* (April 1998), pp. 324-331. A detailed discussion of electrical units, from basic physical laws to commercial calibrations.

Notes and Ideas

*Class time spent on material: Estimated:*_______ *Actual:*_________

Related laboratory activities:

Demonstration materials:

Notes for next time:

Chapter 21: Electric Current and Direct-Current Circuits

Outline

Summary

Basic concepts of DC circuits are covered in Chapter 21. Current, Ohm's law, and power are introduced and applied first to a simple circuit consisting of a battery and a single resistor. Circuits with combinations of resistors in series and parallel are analyzed next, followed by circuits containing combinations of capacitors. Finally, *RC* circuits are considered. At the end of the chapter, devices for measuring current and potential difference in a circuit are discussed.

Major Concepts

By the end of the chapter, students should understand each of the following and be able to demonstrate their understanding in problem applications as well as in conceptual situations.

- Current
  - Definition
  - Batteries
  - Simple circuits
  - Conventional current
- Ohm's law and resistors
  - Resistance and resistivity
  - Ohmic and non-ohmic devices
- DC circuits
  - Energy and power
  - Resistors in series and parallel
  - Kirchhoff's rules
  - Capacitors in series and parallel
  - *RC* circuits
- Ammeters and voltmeters

Teaching Suggestions and Demonstrations

Most people have thrown balls and watched objects fall, so they enter a study of mechanics with everyday experience and observation to build on. Many of your students, however, will not have comparable experiences with circuits. What happens in the wires when a light switch is turned on is invisible to us, so applying common sense and intuition to circuits is difficult. (This may actually be advantageous. Because of friction and air resistance, our "common sense" applied to mechanics is often misleading!) Don't assume previous knowledge, and be sure to use plenty of Conceptual Checkpoints and other thought questions to help students develop understanding of the material. Analogies, like those introduced early in the chapter, are particularly helpful.

Sections 21-1 through 21-3

Begin your presentation by drawing a **simple circuit** consisting of a battery and a lightbulb. Draw both a sketch, showing wires from the two battery terminals connected to the bulb, and a circuit diagram, using the symbols for a battery and a resistor. The **battery** maintains a potential difference between its terminals. If a path is available, electrons will flow from one terminal to the other. The flow of electrons is the **current**. Be sure to differentiate between the actual direction the electrons move and the direction of the **conventional current**, as shown in Figure 21-4. Emphasize that **electromotive force** is not a force at all, but rather a potential difference.

➲ **DEMO 21-1** Bring in three or more common household **batteries** (such as D-cells), a bulb, connecting wires, and a voltmeter. (Although voltmeters aren't discussed until the end of this chapter, you can briefly introduce them here.) Place the batteries in series and then in parallel, testing the voltage across the group each time. (Have students predict the results before you actually perform the demonstration.) Then, connect the batteries (first in parallel and then in series) to the bulb and break the circuit at different points each time. Again, have students predict the results. The bulb goes out no matter where you break the circuit when all the elements are in series, but not when they are all in parallel.

Using a **mechanical analogy for electric circuits**, such as the flow of water illustrated in Figures 21-1 and 21-3, is very helpful to students. The battery is a pump or person raising the water, the current is the flow of the water, and the resistors are water wheels or restricted pipes. Establish this analogy early in the chapter and you will be able to use it in the discussion of resistors in series and parallel later on.

Make sure your students understand the difference between **resistance** and **resistivity**. Example 21-2 provides practice calculating resistance and also justifies the fact that we often ignore the resistance in conducting wires in a circuit. Show samples of color-banded **resistors** used in circuits, but also point out that lightbulbs and toaster ovens act as resistors, since they dissipate electrical energy.

Use the water-flow analogy to help students derive the relationships expressed by **Ohm's law**. If you keep the pump (voltage) constant but use a narrower pipe (increase the resistance), what happens to the water flow (current)? It decreases. If you keep the pipe size (resistance) the same but turn the pump (voltage) up, what happens to the water flow (current)? It increases. Remind students that Ohm's law is not really a law, but rather a relationship that applies to "ohmic" materials.

➲ **DEMO 21-2** Very simple circuits can be used to demonstrate **Ohm's law**. Bring in a power supply, two resistors with very different resistances, an ammeter, and the necessary connecting wires or circuit board. Put one resistor in series with the power supply and ammeter. Read the current for varying values of voltage. Then, replace the resistor and compare the values of current for the same voltages. In the first case, you increase V and examine I with R constant; then you change R and notice the change in I with V constant.

Equations 21-4, 21-5 and 21-6 each express **electrical power** in terms of two of the three variables in Ohm's law (current, voltage and resistance). Remind students again that a **kilowatt-hour** is a unit of energy, not power.

➲ DEMO 21-3 Conceptual Checkpoint 21-2 helps students relate **power** to current, voltage, and resistance. Set it up as a demonstration by including an ammeter and a voltmeter in the circuit and have students calculate the power dissipated by different lightbulbs. (See Resource Information.) This demonstration also provides a nice opportunity to emphasize that a given battery or power supply with constant voltage will not always provide the same current. How much current is drawn from the battery depends on the resistance in the circuit.

➲ DEMO 21-4 The Real World Physics application of the "battery check" meter discussed following Example 21-3 is a clever use of **resistance heating**. Bring in a meter and demonstrate it using both a fully charged and a very weak battery.

Sections 21-4 and 21-5

Once you have introduced the concept of **equivalent resistance for a group of resistors**, derive the relationships specifically for resistors in **series** and in **parallel**. Equivalent resistance of a group of resistors in series will be fairly intuitive for students; add more resistors and you get more resistance. They will likely have more trouble with equivalent resistance of resistors in parallel, when the more resistors you add, the lower the overall resistance becomes. Return to the water analogy to help explain this concept. The short circuit shown in Figure 21-9 is also helpful; if one branch has zero resistance all the current will flow through that path. Be sure to do Examples 21-5 and 21-6 or others like them and compare details of the circuits using the same three resistors in series and in parallel.

➲ DEMO 21-5 Try Conceptual Checkpoint 21-3, which compares identical light bulbs in series and in parallel, as a demonstration. Students should predict the results before you actually connect the circuit. For an extension, you can add more lightbulbs or use an ammeter and a voltmeter to quantify the results.

Reinforce these ideas with plenty of conceptual questions. Consider what happens, for instance, when one more resistor is added to either a series circuit or a parallel circuit. Does the equivalent resistance increase or decrease? What about the current coming from the battery or the current through an individual resistor? What happens to the voltage drop across an individual resistor? You can also briefly discuss **household wiring** at this point. Consider the problems that would be encountered if appliances were connected in series! On the other hand, parallel wiring necessitates fuses or circuit breakers, since adding more branches could increase the total current to dangerous levels.

For **combination circuits**, encourage students to actually sketch the intermediate steps, as is done in Figure 21-10, and to use subscripts on Ohm's law. $V = IR$ is not valid if V refers to the voltage of the battery but I is the current through just one branch. Writing $V_1 = I_1R_1$ or $V_s = I_sR_s$, where *1* refers to an individual resistor and *s* to a subset of the resistors in series, helps students apply Ohm's law correctly.

Figures 21-14 and 21-15 demonstrate **Kirchhoff's rules** applied to a fairly simple circuit. Students can use Ohm's law and equivalent resistance to verify their results. Then move on to a problem such as Active Example 21-2, in which Kirchhoff's rules are necessary. Students tend to have trouble knowing which sign to use for voltages and recognizing the difference between current direction and loop direction. Emphasize that the current direction is your best guess about what is physically happening in the circuit, and the loop is just a way to traverse a part of the circuit, ending up where you started. Use the water analogy again: in a canoe, you can drift downstream and then paddle back upstream to return to your camping spot, but no matter which way *you* are going, the *river* is flowing downstream. As you drift downstream, you lose potential energy, since you are losing elevation. Likewise, traversing a resistor in the direction of the current also corresponds to a potential drop.

➲ DEMO 21-6 Build a **multiloop circuit** with two or more batteries or power supplies and various resistors. Use an ammeter and a voltmeter to demonstrate the junction theorem and the loop theorem.

Sections 21-6 and 21-7

The equations for **equivalent capacitance for capacitors in series and in parallel** can be derived in the same way that the equations for equivalent resistance were derived in Section 21-4. In the case of capacitors, however, a series combination results in an equivalent capacitance that is lower than the capacitance of any one of the individual capacitors, and a parallel arrangement results in a higher capacitance. Remind the students of the physical meaning of capacitance to clarify why this behavior is different from that of resistors.

In an ***RC* circuit**, the rate at which the charge on the capacitor changes depends on the amount of charge itself, hence the exponential relationship. You may need to do a quick review of logarithms in this section. Direct students to the summary of the characteristics of the *RC* circuit near the end of Section 21-7 and discuss what happens to both the current and the charge on the plates of the capacitor during charging and discharging. Windshield wipers and heart pacemakers are two of the applications illustrated.

➲ DEMO 21-7 Set up an ***RC* circuit** with an ammeter in series to demonstrate the time dependence of the charging of a capacitor. Conceptual Checkpoint 21-4 can also be demonstrated. After the capacitor is fully charged, open the switch again and watch what happens to the current as the capacitor discharges through the resistors.

*Section 21-8

If your course has a laboratory component, students will need to be familiar with the operation of **ammeters and voltmeters**. The key differences concern the amount of internal resistance in each and the way each is placed in a circuit. In particular, emphasize that an ammeter, which has a very low internal resistance, must be placed in series with other circuit elements. If placed in parallel, it creates a short circuit that can result in dangerously high currents.

Resource Information

Transparencies

245. Figure 21-1	Water flow as analogy for electric current
246. Figure 21-2	The flashlight: A simple electrical circuit
Figure 21-3	A mechanical analog to the flashlight circuit
247. Figure 21-4	Direction of current and electron flow
248. Figure 21-6	Resistors in series
249. Figure 21-8	Resistors in parallel
250. Figure 21-9	A short circuit
251. Figure 21-10	Analyzing a complex circuit of resistors
252. Figure 21-11	Kirchhoff's junction rule
253. Figure 21-13	Kirchhoff's loop rule

254.	Figure 21-14	Analyzing a simple circuit
	Figure 21-15	Using loops to analyze a circuit
255.	Active Example 21-2	Two Loops, Two Batteries
256.	Figure 21-16	Capacitors in parallel
257.	Figure 21-17	Capacitors in series
258.	Figure 21-18	A typical *RC* circuit
259.	Figure 21-19	Charge versus time for an *RC* circuit
	Figure 21-20	Current versus time in an *RC* circuit
260.	Figure 21-22	Measuring the current in a circuit
261.	Figure 21-23	Measuring the voltage in a circuit

Physlet® Illustrations

21-1 Batteries in Parallel and Series
21-2 Ohm's Law
21-3 Resistivity & Res.
21-4 Resistors in Series and Parallel
21-5 Kirchoff's Law
21-6 RC Circuit
21-7 Capacitors in Parallel and Series

Suggested Readings

Alper, J., "The Battery: Not Yet a Terminal Case," *Science* (17 May 2002), v. 296, pp. 1224-1226. An interesting article on new developments in battery technology.

Amengual, A., "The Intriguing Properties of the Equivalent Resistances of *n* Equal Resistors Combined in Series and in Parallel," *American Journal of Physics* (February 2000), pp. 175-179. An examination of many possible combinations of *n* equal resistors in series and/or in parallel.

Chang, A., "Resistance of a Perfect Wire," *Nature* (3 May 2001), v. 411, pp. 39-40. An interesting discussion of the resistance in electrical contacts.

Cho, A., "New Observations Give Stripes Theory a Lift," *Science* (15 March 2002), v. 295. pp. 1992-1993. A short article on the stripes theory for high-temperature superconductors.

Derman, S. and Goykadosh, A., "A Pencil-and-Tape Electricity Experiment," *The Physics Teacher* (October 1999), pp. 400-402. A simple way of verifying the addition rules for resistors in series and parallel.

Denardo, B., Earwood, J. and Sazonova, V., "Experiments with Electrical Resistive Networks," *American Journal of Physics* (November 1999), pp. 981-986. Two large networks of resistors for use in lecture demonstrations or lab.

Fisher, K., "Conjugate Relationships in Basic Electricity," *The Physics Teacher* (November 1999), p. 458. Help with new electricity vocabulary and relationships.

Hart, F., "Computer-Based Experiments to Measure RC," *The Physics Teacher* (March 2000), pp. 176-177. Computer data acquisition using Vernier equipment.

Henry, D., "Resource Letter TE-1: Teaching Electronics," *American Journal of Physics* (January 2002), pp.14-23. An examination of the role and content of electronics courses with an excellent list of resources.

Herman, R., "An Introduction to Electrical Resistivity in Geophysics," *American Journal of Physics* (September 2001), pp.943-952. A nice application of an introductory physics topic to geophysics, including description of a resistivity apparatus.

Service, R., "MgB2 Trades Performance for a Shot at the Real World," *Science* (1 February 2002), v. 295, pp. 786-788. An interesting article on practical superconductors.

Service, R., "Shrinking Fuel Cells Promise Power in Your Pocket," *Science* (17 May 2002), v. 296, pp. 1222-1224. An interesting article on new micro fuel cells.

Voss, D., "Perplexing Compounds Rejoin the Club," *Science* (25 January 2002), v. 295, p. 604. A short article on high temperature superconductors.

Vreeland, P., "Analyzing Simple Circuits," *The Physics Teacher* (February 2002), pp. 99-100. An example of a graphic organizer that helps students correctly apply Ohm's law to series and parallel circuits.

Materials

Ammeters and voltmeters are available from a variety of sources. One voltmeter/ammeter/ohmmeter combination is available from Sargent-Welch, model number WLS30710-10.

Notes and Ideas

*Class time spent on material: Estimated:*_______ *Actual:*______________

Related laboratory activities:

Demonstration materials:

Notes for next time:

Chapter 22: Magnetism

Outline

Summary

Chapter 22 introduces magnetic fields and magnetic forces. A moving charge will experience a force in a magnetic field if a component of its velocity is perpendicular to the field. Thus, a current-carrying wire may experience a force and a current-carrying loop may experience a torque in a magnetic field as well. Magnetic fields arise from moving charges. The specific fields due to current-carrying straight wires, wire loops and solenoids are examined. Magnetism in matter and the Earth's magnetic field are also discussed.

Major Concepts

By the end of the chapter, students should understand each of the following and be able to demonstrate their understanding in problem applications as well as in conceptual situations.

- Magnets and the magnetic field
  - North and south poles
  - Field lines
  - Earth's magnetic field
- Magnetic force
  - Moving charged particle
  - Magnetic force right-hand rule
  - Current-carrying wire
  - Torque on a current-carrying loop
- Magnetic field
  - Magnetic field right-hand rule
  - Long, straight wire
  - Current loop
  - Solenoid
- Magnetism in matter

Teaching Suggestions and Demonstrations

One main point of Chapter 22 is that electricity and magnetism are intimately connected. The earlier sections deal with magnetic forces on moving electric charges. In the later sections, students find that magnetic fields *arise* from moving electric charges. None of this is intuitively obvious. Use plenty of conceptual questions and examples to check and reinforce student understanding.

Section 22-1

In studying the electric field, students found that the field direction could be determined by the direction of the force on a positive test particle. In a similar way, the direction of a **magnetic field** can be found using a compass as a "test particle." Point out some differences between the two types of field; electric field lines begin and end on charges but magnetic field lines form closed loops. Although positive and negative electric charges may exist individually, north and south magnetic poles cannot, as illustrated in Figure 22-2.

➲ **DEMO 22-1** Bring in a bar magnet and use a small compass to "test" the **magnetic field** around it. After students have a feel for the shape of the field, place a sheet of paper over the magnet and sprinkle iron filings on top, as shown in Figure 22-3. Alternatively, you can purchase a "magnetic field demonstrator" which consists of a transparent box filled with iron filings suspended in a liquid, with a hole in the center in which a magnet can be inserted. Students will be able to see the filings line up in three dimensions. (See Resource Information.)

Atmospheric auroral displays are caused by the interaction of particles from the Sun with particles in the Earth's atmosphere. The reason these beautiful light shows occur at high latitudes (and are therefore called northern or southern lights) is that the solar particles are directed toward the poles by the **magnetic field of the Earth**, as discussed later in Section 22-3. (See Resource Information.)

Sections 22-2 and 22-3

You can begin the somewhat bewildering topic of **magnetic force on moving charges** by asking students what conditions must be met for a particle to experience a force in the other fields encountered so far. All particles experience a force in a gravitational field, but only charged particles do in an electric field. In the case of the magnetic field, the particle must not only be charged, but must also be moving with a component of its velocity perpendicular to the field. The magnitude of the magnetic force is actually defined by the force, charge, speed, and angle.

After introducing the **magnetic force right-hand rule** and the conventions for drawing vectors into and out of the page, draw lots of examples on the board or overhead and have the class practice determining the direction of the magnetic force. Vary your examples as much as possible. Draw the fields pointing in or out as well as in the plane of the board. Be sure to include a few examples where the velocity and field are parallel or anti-parallel, in which case the force is zero. Also toss in a few negative particles to remind students that the force in this case is opposite the result given by the right-hand rule. Conceptual Checkpoint 22-2 is a good one to do together.

The discussion of electric versus magnetic forces provides a nice review of work and the work-energy theorem. Once students are convinced that the magnetic force does no work and therefore cannot change the speed of the particle, ask them about the velocity of the particle. The magnetic force changes the direction of the velocity vector, but not the magnitude. The mass spectrometer and the electromagnetic flowmeter discussed in Section 22-3 are both good examples of applications that will be of particular interest to your biology majors and students in the pre-health professions.

➲ **DEMO 22-2** Show students photographs of particle tracks from bubble, cloud or spark chambers and point out the spiraling paths. Relate them to Equation 22-3 for the **radius of the orbit of a moving charged particle in a magnetic field** and discuss the various parameters that can be determined from the experimental data. (See Resource Information.)

Sections 22-4 and 22-5

Students now know that a single charged particle moving in a magnetic field may experience a force. Next, ask them to consider a group of charged particles moving in the same direction and they will see that a **current-carrying wire in a magnetic field** will also experience a force.

➲ **DEMO 22-3** To demonstrate the **force on a current-carrying wire in a magnetic field**, place a wire connected to a battery or power supply between the poles of a (strong!) horseshoe magnet as shown in Conceptual Checkpoint 22-4. When current is flowing, the wire will be deflected. Reverse the current direction and notice that the deflection also reverses. Turn the current off to demonstrate that the deflection really is due to the moving charges. (Be sure to experiment with this demonstration ahead of time so you will know how much current is necessary to produce a noticeable result for your magnet.) The students can use the right-hand rule and their observations to determine which end of the magnet is the north pole.

➲ **DEMO 22-4** The easiest way to introduce **torque on a current-carrying loop** is through a demonstration. A wire coat hanger bent into a rectangle or four rulers taped together work fine. Establish a direction in the classroom for an imaginary magnetic field. Place an easily visible marker of some kind on the rectangular loop to indicate current direction. Then hold the loop in the "magnetic field" with the plane of the loop parallel to the field and ask students to use the right-hand rule to determine the direction of the force on each of the four sides. Try different orientations to illustrate maximum torque and zero torque. Note that if the torque causes the loop to "overshoot" the zero position, the new torque attempts to return the loop to the zero position instead of continuing a circular motion. You can discuss the need for a commutator to change current direction in order to produce continuous motion, as in a motor.

Sections 22-6 and 22-7

The previous sections have dealt with the magnetic force on moving charges. These next two sections discuss the fact that magnetic fields are actually created by moving charges, or current. The **magnetic field right-hand rule** gives the relationship between current direction and resulting field. It is sometimes referred to as the right-hand *wire* rule in order to avoid confusion with the magnetic force right-hand rule.

Point out to students that **moving charges create a magnetic field**, but that equations for only three specific configurations (long straight wire, current loop, and solenoid) will be discussed. Although the formula for the field in the center of a current loop of radius r is not derived, you can compare it to the formula for the field the same distance r away from a long straight wire. It will make sense to students that the field in the center of the loop is stronger, since the point in question is surrounded by current on all sides. When discussing the solenoid, point out why the magnetic field *outside* is negligible.

The **force between two current-carrying wires** is a nice summary of the main points of the chapter. Magnetic fields are produced by moving charges; moving charges in magnetic fields experience forces. Both right-hand rules are necessary to illustrate that two wires carrying currents in the same direction will attract each other whereas two wires with currents in opposite directions will repel. Do this together with students and encourage them to try it again on their own for more practice with the direction rules.

➲ DEMO 22-5 Set up two **parallel wires** and run currents in the same and in opposite directions to illustrate the attractive and repulsive forces that result.

Section 22-8

The details of ferromagnetism, paramagnetism and diamagnetism can be skipped if necessary due to time constraints. It is a good idea, however, to mention the atomic level structure of "permanent magnets" to show that even these owe their magnetic fields to moving charges.

Resource Information

Transparencies

262.	Figure 22-1	The force between two bar magnets
	Figure 22-2	Magnets always have two poles
263.	Figure 22-4	Magnetic field lines for a bar magnet
264.	Figure 22-6	Magnetic field of the Earth
265.	Figure 22-7	The magnetic force on a moving charged particle
266.	Figure 22-8	The magnetic force right-hand rule
267.	Figure 22-9	The magnetic force for positive and negative charges
268.	Figure 22-10	Differences between motion in electric and magnetic fields
269.	Figure 22-12	Circular motion in a magnetic field
270.	Figure 22-15	The magnetic force on a current carrying wire
271.	Figure 22-17	Magnetic torque on a current loop
272.	Figure 22-19	The magnetic field of a current-carrying wire
	Figure 22-20	The magnetic field right-hand rule
273.	Figure 22-23	The magnetic force between current-carrying wires
274.	Figure 22-27	The solenoid

Physlet® Illustrations

22-1 The Field of a Bar Magnet
22-2 The Magnetic Field Around a Current-Carrying Wire

Suggested Readings

Baker, B., "Demonstrating Forces Between Parallel Wires," *The Physics Teacher* (May 2000), p. 299. A dramatic illustration of the repulsion or attraction between parallel current-carrying wires.

Banerjee, S., "When the Compass Stopped Reversing Its Poles," *Science* (2 March 2001), v. 291, pp. 1714-1715. An accessible, short article on the long-term behavior of the Earth's magnetic field.

Bichsel, E., Wilson, B. and Geerts, W., "Magnetic Domains of Floppy Disks and Phone Cards Using Toner Fluid," *The Physics Teacher* (March 2002), pp. 150-153. A conceptual demonstration using dry and liquid toner to find the magnetic poles at the surfaces of refrigerator magnets and magnetic recording media.

Brown, R., "Demonstrating the Meissner Effect *and* Persistent Current," *The Physics Teacher* (March 2000), pp. 168-169. A demonstration involving levitation of a magnet over a superconductor.

Cavicchi, E., "Experiences with the Magnetism of Conducting Loops: Historical Instruments, Experimental Replications, and Productive Confusions," *American Journal of Physics* (February 2003), pp. 156–167. An interesting study of nineteenth century lab work on electromagnetism.

Chia, C. and Wang, Y., "The Magnetic Field Along the Axis of a Long Finite Solenoid," *The Physics Teacher* (May 2002), pp. 288-289. Investigates the field of a finite solenoid with a multimeter and a pick-up coil.

Erlichson, H., "The Experiments of Biot and Savart Concerning the Force Exerted by a Current on a Magnetic Needle," *American Journal of Physics* (May 1998), pp. 385-391. Descriptions of many of the historically important experiments relating electricity and magnetism.

Gibson, F. and MacInnes, I., "Symmetry in Electromagnetism: A New Magnetic Needle," *The Physics Teacher* (May 2000), pp. 316-317. Simple experiments to demonstrate the relationship between electricity and magnetism.

Grünberg, P., "Layered Magnetic Structures: History, Highlights, Applications," *Physics Today* (May 2001), pp. 31-37. Essay on thin magnetic films with unique properties and their many applications, included electronic data storage.

Schmidt, M., "Investigating Refrigerator Magnets," *The Physics Teacher* (April 2000), pp. 248-249. A nice complement to Walker's discussion of refrigerator magnets.

Sinacore, J. and Graf, E., "Precision Laboratory Apparatus for the Study of Magnetic Forces," *The Physics Teacher* (May 2000), pp. 296-298. A rugged and precise apparatus for introductory magnetism labs.

Stewart, G., "Measuring Earth's Magnetic Field Simply," *The Physics Teacher* (February 2000), pp. 113-114. An easy and popular activity to measure the Earth's magnetic field.

Van Heuvelen, A., Allen, L. and Mihas, P., "Experiment Problems for Electricity and Magnetism," *The Physics Teacher* (November 1999), pp. 482-485. Real life laboratory problems for introductory electricity and magnetism.

Welsh, G., "Magnetic Therapy in Physics?," *The Physics Teacher* (March 2000), pp. 181-182. A critical thinking and writing assignment on magnetic therapy.

Wright, K., "Seeing the Light," *Discover* (July 2000), pp. 51-57. A very complete discussion of auroras, with great pictures.

Materials

"Aurora: Rivers of Light in the Sky," videotape, SkyRiver Films, Anchorage, Alaska, 1-800-248-WILD.

The Fisher Science "Deluxe Magnetic Field Apparatus" (item # S43059-1) allows students to see iron filings align with the magnetic field in three dimensions.

Bubble chamber photographs can be found on the web at the Fermi National Accelerator Laboratory website, http://www.fnal.gov/. Go to the search option and type in "bubble chamber photographs". There is a section with an example photograph and sample questions that can be answered by simple analysis of the picture.

Notes and Ideas

Class time spent on material: Estimated:________ Actual:______________

Related laboratory activities:

Demonstration materials:

Notes for next time:

Chapter 23: Magnetic Flux and Faraday's Law of Induction

Outline

Summary

Magnetic fields arise from moving electric charges, as shown in Chapter 22. Chapter 23 looks at the reverse relationship: electric fields are induced by changing magnetic fields. Faraday's law, which quantifies the electromotive force induced by a changing magnetic flux, and Lenz's law, which gives the direction of the induced current, are both described in detail. Electrical and mechanical energy are also discussed. Generators and motors transform between the two types of energy. Inductors store energy in a magnetic field, and transformers convert power at one voltage and current to power at another voltage and current.

Major Concepts

By the end of the chapter, students should understand each of the following and be able to demonstrate their understanding in problem applications as well as in conceptual situations.

- Magnetic flux
- Induced emf
  - Faraday's law
  - Lenz's law
  - Motional emf
- Mechanical work and electrical energy
  - Generators
  - Motors
- Inductance
  - Inductors and *RL* circuits
  - Energy stored in a magnetic field
  - Transformers

Teaching Suggestions and Demonstrations

Electromagnetic induction almost looks like magic. At first encounter it is pretty impressive that merely waving a magnet around in the vicinity of a coil or twisting a loop in an external magnetic field can produce a current. This chapter is fairly long, but don't skimp on the time you spend on Faraday's law and Lenz's law. The remaining sections are essentially extensions or applications of these principles, so ensuring a firm understanding of the early part of the chapter will allow you to move more quickly through the later sections. Mention practical applications of electromagnetic induction that students encounter in their everyday lives, like credit card readers and magnetic tape recordings.

Sections 23-1 through 23-4

➲ DEMO 23-1 A nice way to begin the discussion of **induced emf** is through the demonstration shown in the photograph in Section 23-1. Connect a solenoid to a sensitive ammeter. Insert a bar magnet into the solenoid while watching the ammeter needle. Also, hold the magnet still inside the coil and then remove it. Repeat, varying the speed of the magnet. As an extension, use a second coil attached to a power supply as the source of the magnetic field. In this case, students will watch the ammeter as the current in the coil is turned on, kept on, and then turned off. Ask students what caused the current registered by the ammeter. Since there was no induced current when the magnet was still or the current in the primary coil was constant, they should recognize that a magnetic field alone doesn't induce a current. Clearly some sort of *change* is necessary. This leads into a discussion of magnetic flux.

The water passing through a fishing net in a stream is a useful analogy for **magnetic flux**. How much water passes through the net depends on the size of the net, the orientation of the net with respect to the flowing water, and how fast the water is moving. Likewise, magnetic flux depends on the area of interest, the angle its normal makes with the magnetic field direction, and the magnitude of the magnetic field. It is worth spending some time on the concept of the **normal to the surface** and the angle between the normal to the surface and the magnetic field. Remind students to use their common sense. If the plane of a loop is perpendicular to the magnetic field, then clearly the flux is a maximum. They'll be tempted to hear "perpendicular" and let $\theta = 90^o$, but since $\cos 90^o = 0$, they should catch the mistake.

To set the stage for **Faraday's law**, try Conceptual Checkpoint 23-1 regarding changing magnetic flux through a loop. Emphasize that a changing flux doesn't necessarily require a changing field. Show examples of changing area, as in motional emf, as well as changing field. Point out to students that a third way to change the magnetic flux through a loop is to change the angle between the loop and the field. A rotating loop in an external magnetic field is the basis of a generator, which will be discussed in a later section.

Do plenty of conceptual examples for **Lenz's law**. Students catch on to the idea of "opposition" fairly easily, but can get confused about just what is being opposed. Include examples in which the original field decreases to emphasize that the induced current creates an induced magnetic field that opposes the *change*, but doesn't necessarily oppose the original field.

Sections 23-5 and 23-6

A moving rod has mechanical energy. In the case of motional emf, as shown in Figure 23-13, this **mechanical energy is converted into electrical energy**. The motion of the rod changes the area of the rectangular loop in the magnetic field, which, consistent with Faraday's law, causes an induced emf. Another way to convert mechanical energy into electrical energy is to rotate a conducting loop in a magnetic field. An emf is induced since the magnetic flux through the loop is changing.

➲ DEMO 23-2 Pull out the bent coat hanger or four rulers taped into a rectangle that you used in the discussion of torque on a current loop in Chapter 22. Now this fancy piece of apparatus will become a **generator**. Establish the direction of the imaginary external magnetic field and rotate the rectangle. Point out to students that you are supplying mechanical energy by turning the loop. Faraday's law states that a current will be induced. To establish the alternating nature of the current, have students apply the magnetic force right-hand rule to the separate sections of the loop for the different positions in the rotation.

The demonstration above convinces students that the current direction alternates. Although calculus is handy for deriving that the alternation is sinusoidal, it isn't actually necessary. Apply the formula for motional emf, $\varepsilon = vBl$, but point out that only the component of the velocity perpendicular to the magnetic field contributes, so replace v with $v\sin\theta$.

Return to the Chapter 22 discussion of torque on a current loop in a magnetic field and contrast **electric motors** with generators. The former convert electrical energy to mechanical work and the latter convert mechanical work to electrical energy. Demonstration motors/generators are available from several scientific equipment companies. (See Resource Information.)

Sections 23-7 through 23-9

Mutual inductance and **self inductance** are further applications of Faraday's and Lenz's laws. Both inductors and capacitors will be used in the circuits discussed in the next chapter, so it is a good idea to draw comparisons between them now. **Inductance** is a property of an **inductor** based upon its physical characteristics, just as capacitance depends only on the characteristics of the capacitor and not on the circuit in which the capacitor is placed. When used in a circuit, both store energy in fields—the inductor in a magnetic field and the capacitor in the electric field between its plates. The characteristic decay times for each type of circuit can also be compared.

➲ DEMO 23-3 Demonstrate **back emf** in an inductor in an *RL* circuit using a function generator on square wave mode as the power supply. Use an oscilloscope to examine the voltage across the inductor. (See Resource Information.) The square wave simulates a DC circuit with a switch repeatedly opened and closed. Each time the switch opens or closes, the back emf jumps to a maximum value because the current is changing and thus the magnetic flux through the coil is changing. In between, as the current becomes established at a constant level, the back emf drops to zero because the magnetic flux in the coil becomes constant.

Section 23-10

A real advantage of alternating current over direct current is the ability of **transformers** to step up or step down the voltage of an AC source. Transformers don't work with direct current because, as Faraday's law states, there must be a changing magnetic flux in order to induce an emf in the secondary coil. Remind students that conservation of energy requires that the power remain constant, so if voltage is stepped up by a factor of 10, current must go down by the same factor. Since losses in power lines depend on the square of the current, stepping current down by a factor of 10 reduces power lost to heating by a factor of 100.

Resource Information

Transparencies

275. Figure 23-1 — Magnetic induction
276. Figure 23-2 — Induced current produced by a moving magnet
277. Figure 23-3 — The magnetic flux through a loop

278.	Conceptual Checkpoint 23-1	
279.	Figure 23-4	A dynamic microphone
	Figure 23-5	The pickup on an electric guitar
	Figure 23-6	Magnetic tape recording
280.	Figure 23-8	Applying Lenz's law to a magnet moving toward and away from a current loop
281.	Figure 23-10	Motional emf
	Figure 23-11	Determining the direction of an induced current
282.	Figure 23-12	Eddy currents
283.	Figure 23-13	Force and induced current
284.	Figure 23-14	An electric generator
	Figure 23-16	A simple electric motor
285.	Figure 23-17	A changing current in an inductor
286.	Figure 23-19	An *RL* circuit
	Figure 23-20	Current as a function of time in an *RL* circuit
287.	Figure 23-22	The basic elements of a transformer

Physlet® Illustrations

23-1 Faraday's Law

Suggested Readings

Baldwin, N., *Edison: Inventing the Century*, Hyperion, New York, 1995. An excellent biography of Edison. Chapter 17 discusses the debate on whether to use alternating or direct current as the electrical standard.

Higbie, J., "A Novel Approach to Faraday's Law," *The American Journal of Physics* (December 1997), p. 1211. Uses a hypothesis of magnetic monopoles to arrive at Faraday's law.

Ivanov, D., "Another Way to Demonstrate Lenz's Law," *The Physics Teacher* (January 2000), pp. 48-49. Uses falling magnets to illustrate Lenz's law.

Kingman, K. and Popescu, S., "Motional EMF Demonstration Experiment," *The Physics Teacher* (March 2001), pp. 142-144. A very visual demonstration of electromagnetic induction.

Kingman, R., Rowland, S. and Popescu, S., "An Experimental Observation of Faraday's Law of Induction," *American Journal of Physics* (June 2002), pp. 595-598. Description of an experimental apparatus demonstrating Faraday's law.

Lan, B. and Lim, J., "Michael Faraday: Prince of Lecturers in Victorian England," *The Physics Teacher* (January 2001), pp. 32-35. An informative historical article on Faraday.

Sawiki, C., "A Lenz's Law Experiment Revisited," *The Physics Teacher* (October 2000), pp. 439-441. Magnetic braking in a real-world situation.

Tanner, P., Loebach, J., Cook, J. and Hallen., H., "A Pulsed Jumping Ring Apparatus for Demonstration of Lenz's Law," *American Journal of Physics* (August 2001), pp. 911-916. Provides instructions for construction and use of a good demonstration of Lenz's law.

Materials

One demonstration motor is available from Fisher Science Education, model number S43580.

Function generators and oscilloscopes are available from a wide range of manufacturers.

An eddy current demonstrator is available from Fisher Science Education, model number S83019. Klinger Educational Products has a demonstration of an eddy-current brake, P3.4.4.

Notes and Ideas

Class time spent on material: Estimated:______ *Actual:____________*

Related laboratory activities:

Demonstration materials:

Notes for next time:

Chapter 24: Alternating-Current Circuits

Outline

Summary

The generator introduced in Chapter 23 produces alternating, sinusoidal current and voltage. Chapter 24 examines circuits with such an AC generator and a combination of resistors, capacitors and inductors. A generalized notion of resistance, the impedance, is used in analyzing these *RLC* circuits. The phase relationships among the voltage drops across the three different elements and the current are also discussed in detail. Finally, resonance in an *RLC* circuit and its applications are introduced.

Major Concepts

By the end of the chapter, students should understand each of the following and be able to demonstrate their understanding in problem applications as well as in conceptual situations.

- Alternating current and voltage
  - Phasors and phasor diagrams
  - RMS values
- Motional EMF
- Capacitors in AC circuits
  - Capacitive reactance
  - Phase relation between current and voltage
- Inductors in AC circuits
  - Inductive reactance
  - Phase relation between current and voltage
- *RLC* circuits
  - Impedance
  - Phase angle
  - Power factor
  - Resonance

Teaching Suggestions and Demonstrations

Diagrams and graphs are very helpful in understanding *RLC* circuits. Use them liberally throughout the chapter and encourage students to make sketches as they work problems. The many new terms introduced in this chapter can be confusing. Spend time clarifying terms such as reactance and impedance and show students how they relate to familiar properties such as resistance.

Section 24-1

Root mean square values of alternating current and voltage are often difficult for students to understand at first. The typical household plug provides an rms voltage of 120 V, which corresponds to a maximum voltage of 170 V (see Exercise 24-1). The 170 V figure is in a sense not very meaningful; the voltage only reaches that value for brief moments in the cycle. The wall socket *effectively* supplies 120 V, which means that the power dissipated is equivalent to the power that would be dissipated in the same appliance if it were connected to a 120 V direct current source. In fact, rms values for current and voltage are sometimes called effective values.

Phasors and **phasor diagrams** are also very tricky. Introduce them here and convince students that the angular speed ω of the phasor's rotation about the origin is the same as the angular frequency of the sinusoidally alternating voltage. Since the current and voltage of the resistor in an AC circuit are in phase, phasor diagrams aren't really that useful in analyzing circuits that have only resistors. However, they're extremely helpful when capacitors and inductors are in the circuit, so it's best to introduce them on the easier case and let students become familiar with them.

Sections 24-2 through 24-5

In order to give students an idea of the goal of these next sections, draw a circuit with a battery and three resistors in series and then a separate circuit with **an AC generator and a resistor, an inductor and a capacitor in series**. Ask students how the voltage drops across the three resistors in the battery circuit are related to the voltage of the battery. They will respond that the resistor voltage drops add up to the total gain from the battery, or $V = V_1 + V_2 + V_3$. A reasonable guess for the AC circuit then would be $V_{gen} = V_R + V_L + V_C$. However, for the AC circuit, the voltages refer to rms values and they do not add up simply to equal the rms voltage of the generator. The reason is that the different sinusoidal voltages are not in phase. The first task, then, is to discover the **phase relationships**.

Students already know that the voltage and current of the **resistor** are in phase and are related by Ohm's law, which applies to both alternating and direct current circuits as long as V and I correspond to the same type of value, for instance, rms or maximum values.

A **capacitor** in an ac circuit is more complicated. Sketch a sine curve to represent the current in the circuit. Plot the corresponding voltage across the plates of the capacitor by discussing the behavior of the charges. For instance, when the voltage across the plates is zero, the current will be a maximum since there is nothing to oppose it. When the voltage across the plates is a maximum, the current will be zero.

Provide a similar exercise for an **inductor** in an ac circuit. Because of Faraday's law, the voltage across the inductor depends on the *rate of change* of the current, so when the current reaches a maximum, the voltage will be zero. These graphs show that the voltage across the inductor leads the current and the voltage across the capacitor lags the current, each by 90°.

For both capacitor and inductor, note that the **reactance** is a measure of the opposition to current flow. The expressions for the reactances are not derived, but it is a good idea to verify with students that they are at least reasonable. First, have students check units to make sure both are in ohms. Then look at the extreme cases of frequency. When frequency equals zero, which corresponds to the direct current case, the capacitive reactance becomes infinite and the inductive reactance drops to zero, both as expected. Repeat for the case of a very large frequency. Figure 24-14 illustrates the **frequency dependence** of the resistance and the reactances. Figure 24-22 illustrates a method for predicting the current at the low-frequency and high-frequency extremes.

Now that the students have an understanding of the **leading and lagging** of the various voltages with respect to the current, they will appreciate the usefulness of a **phasor diagram**, such as the one shown in Figure 24-21. Remind them that all the phasors except the generator voltage remain in fixed relationships with each other. (For example, V_L always leads V_R and I by 90°.) The phasor diagram helps students visualize the **phase angle,** which in turn gives them a feel for the **power** dissipated in the circuit.

➲ DEMO 24-1 Although the rms voltages of the elements in the *RLC* circuit do not add up simply, the **sum of the sine curves** for the voltages of the three elements does represent the sinusoidal voltage of the generator. It is hard to add sine curves free-hand, so prepare some computer plots ahead of time. Graph the curves representing v_L, v_R, and v_C and then show their sum. This will be most effective if the numbers correspond to an example you will be doing in class, such as Example 24-6, so the students can see that the sum not only is another sine curve but also has the expected leading or lagging relationship. If desired, you can show graphs for *RL* and *RC* circuits separately before proceeding to the *RLC* example.

Section 24-6

Introduce **resonance** by asking students how to increase current in a given *RLC* circuit. Maximum current will correspond to the minimum impedance. An examination of the formula for impedance indicates that this will occur when $X_L = X_C$. Alternatively, examine the phasor diagram and note that the greatest power results when V_R is a maximum. The largest value V_R can have is the same as the generator voltage, and again, this results when $X_L = X_C$. Figure 24-24 shows both impedance and current as a function of frequency. The discussion of the *LC* circuit relates resonance in an electrical circuit to resonance of a mechanical system.

➲ DEMO 24-2 Set up an ***RLC* series AC circuit** with a variable frequency generator, an inductor, a resistor and a capacitor. Use an oscilloscope to examine the voltage across the resistor and the generator voltage, displaying both sine curves at once. (Since the resistor voltage is always in phase with the current, it can be used to determine whether the generator voltage is leading or lagging the current.) Begin with frequencies lower than resonance frequency and watch the two curves as you increase to and then beyond the resonance frequency. Students should be able to see both the changing phase relationship and the changing magnitude of V_R. At resonance the two curves will be identical.

Resource Information

Transparencies

288. Figure 24-2	AC voltage and current for a resistor circuit
289. Figure 24-3	Phasor diagram for an ac resistor circuit
290. Figure 24-7	Time variation of voltage, current and power for a capacitor in an ac circuit
291. Figure 24-8	Current and voltage phasors for a capacitor
Figure 24-10	Phasor diagram for an *RC* circuit
292. Figure 24-12	Voltage, current, and power for various phase angles, ϕ
293. Figure 24-14	Frequency variation of the inductive reactance, X_L, the capacitive reactance, X_C, and the resistance, R
294. Figure 24-15	Voltage, current, and power in an ac inductor circuit
295. Figure 24-16	Phasor diagram for an ac inductor circuit
Figure 24-18	Phasor diagram for an *RL* circuit

296. Figure 24-19	RMS currents in *RL* and *RC* circuits as a function of frequency
297. Figure 24-20	An alternating-current *RLC* circuit
Figure 24-21	Phasor diagram for a typical *RLC* circuit
298. Table 24-1	Properties of AC Circuit Elements and Their Combinations
299. Figure 24-22	High-frequency and low-frequency limits of an ac circuit
300. Figure 24-23	Oscillations in an *LC* circuit with no generator
301. Figure 24-24	Resonance in an *RLC* circuit

Physlet® Illustrations

24-1 Capacitors in AC-Circuits
24-2 Inductors in AC-Circuits
24-3 Impedance in an RLC Circuit
24-4 Resonance in an RLC Circuit

Suggested Readings

Backman, P., Murley, C. and Williams, P., "The Driven *RLC* Circuit Experiment," *The Physics Teacher* (October 1999), pp. 424-425. Uses a virtual four-channel oscilloscope installed on a laptop computer to display voltages across all elements in an *RLC* circuit simultaneously.

Cervellati, R. and Soldà, R., "An Alternating Voltage Battery with Two Salt-Water Oscillators," *American Journal of Physics* (May 2001), pp. 543-545. Describes the construction and operation of a battery that reverses polarity.

Fay, S. and Portenga, A., "Hey You! Shut the Refrigerator Door!," *The Physics Teacher* (September 1998), pp. 336-338. A great experiment on power usage by electrical appliances.

Johns, R., "Simplifying AC-Current Measurements," *The Physics Teacher* (May 2001), pp. 314-315. Describes a simple procedure using AC power strips for measuring the current drawn by household appliances.

Oliver, J., "Observing Voltage Phases in *RC*, *RL*, and *RLC* Circuits," *The Physics Teacher* (January 1997), p. 30. A short note on experimental observations of voltage phases.

Ruby, L., "Why DC for Long-Range Power Transmission," *The Physics Teacher* (May 2002), pp. 272-274. A brief history of the debate regarding AC versus DC usage for electrification of the United States, followed by arguments in support of DC.

Silverman, M., "Power, Reaction, and Excitement…in an AC Circuit," *The Physics Teacher* (May 2002), pp. 302-307. Excellent discussion of AC circuits involving fluorescent lamps.

Notes and Ideas

*Class time spent on material: Estimated:*______ *Actual:*______________

Related laboratory activities:

Demonstration materials:

Notes for next time:

Chapter 25: Electromagnetic Waves

Outline

Summary

Chapter 25 brings together concepts from Chapters 19-24. The production and propagation of electromagnetic waves are discussed and the entire electromagnetic spectrum is presented. The transfer of energy and momentum in electromagnetic waves is explained. The chapter ends with a section on the polarization of electromagnetic waves.

Major Concepts

By the end of the chapter, students should understand each of the following and be able to demonstrate their understanding in problem applications as well as in conceptual situations.

- Existence of electromagnetic waves
  - Symmetry arguments
  - Generation of electromagnetic waves
- Propagation of electromagnetic waves
  - Speed
  - Direction of propagation
  - The Doppler effect
- The electromagnetic spectrum
- Energy and momentum in electromagnetic waves
  - Energy density
  - Radiation pressure
- Polarization
  - Transmission axis
  - Law of Malus

Teaching Suggestions and Demonstrations

This is an exciting chapter to teach. The discoveries that electromagnetic waves travel at the speed of light and that light is an electromagnetic wave were astounding to scientists of the time. Try to convey the sense of amazement that must have accompanied this realization.

Sections 25-1 through 25-3

Plan to spend a little class time talking about **Maxwell's symmetry arguments** for the existence of magnetic fields produced by changing electric fields and his theoretical prediction of **electromagnetic waves** that travel at the speed of light. (While this is well-known today, it came as a shock to scientists of the 1800's!) Go through the historical developments sketched at the beginning of Section 25-1 and through the production of an electromagnetic wave by an antenna shown in Figures 25-1, 25-2, and 25-3. Point out that electromagnetic waves are **transverse waves** and that the electric and magnetic fields are perpendicular to each other and in phase. Show the students how to use the right-hand rule to find the **direction of propagation** of an electromagnetic wave. Use Conceptual Checkpoint 25-1 to test understanding. Be sure students are aware that ***all*** **accelerating charges radiate electromagnetic waves.**

Scientists initially postulated the existence of the **ether** as the medium through which electromagnetic waves travel. The story of the search for the ether is quite interesting; it was a major scientific controversy of the nineteenth century. It does not take long to go over in class, and it gives students the sense of the "real" way science develops—lots of people arguing over the interpretation of data! (See Resource Information.)

All electromagnetic waves in a vacuum travel at the **speed of light**. The speed of light is a large number. Give students several real world estimates to help them understand how fast light travels. Work Exercise 25-1 in class. The historical material presented in Section 25-2 on the measurement of the speed of light can be left for the students to read if you are short on time. Do plan to spend time on the **Doppler effect** for electromagnetic waves, even if you just discuss it qualitatively. The applications to police and weather radar (and to astronomy) are fascinating. Work Example 25-2.

The inverse relationship between the **wavelength and frequency** of electromagnetic waves may take the students some time to grasp. Give many reminders (energy increases with frequency and decreases with wavelength) and be consistent about how you graph these quantities. Go through the **electromagnetic spectrum**, giving examples for each region.

➲ **DEMO 25-1** It is a good idea to have a chart of the **electromagnetic spectrum** (see Figure 25-8) to show students as you talk about it. Be sure to point out what a small section of the spectrum is made up of visible light.

Section 25-4

Electromagnetic waves are made up of electric and magnetic fields, each of which has an **energy density.** Therefore, electromagnetic waves have an energy density and can transport energy from one place to another. Work through the calculation of the total energy density of an electromagnetic wave at the beginning of this section and the relation $E = cB$ for the electric and magnetic fields in an electromagnetic wave. Define the **rms values** of the electric and magnetic fields and the **intensity** of the electromagnetic wave. Work Example 25-4 in class to show how intensity decreases with distance.

➲ **DEMO 25-2** Bring in a HeNe laser and do Active Example 25-1 to connect the theory of this section to real objects. Have a student measure the size of the beam as it emerges from the laser. Use the actual value of the diameter of the beam for the calculation.

Most students will find it easy to understand that electromagnetic waves carry energy. The ideas that these waves also carry **momentum** and can produce **radiation pressure** are much less intuitive. Work through the theory carefully and do many examples, including Conceptual Checkpoint 25-2 and Exercise 25-4. Discuss how radiation pressure produces the tail of a comet.

Section 25-5

Begin this section on the **polarization** of light by discussing the orientation of the electric field in an electromagnetic wave. Point out that in most light sources, the light is generated randomly by many atoms. Each bit of light is produced separately and each has its own polarization direction. Such a source is called **unpolarized**, because the combination of all the bits of light has no overall polarization. (See Figure 25-12.) Then you can discuss how a polarizer allows electric fields with certain orientations through and stops others. Each polarizer has a **transmission axis**.

➲ DEMO 25-3 Get a large piece of cardboard (at least 0.5 m square) and cut a slit in it a little wider than a demonstration Slinky®. Use it and the Slinky® to demonstrate the **action of a polarizer**, as shown in Figure 25-13. (Be sure to point out that the direction of the slit in the cardboard corresponds to the transmission axis of the polarizer and is perpendicular to the direction of the long molecular chains.)

➲ DEMO 25-4 It is well worth the cost to purchase several sheets of **polarizing material** large enough to cover the top of an overhead projector. (See Resource Information.) Use the projector as the incandescent light source. Place one of the polarizer sheets over the top. Students will notice a decrease in the intensity of the light. (See Figure 25-15.) Now place a second polarizer on top of the first, with the transmission axes aligned. Rotate the top polarizer slowly through 90 degrees and show the resultant decrease in the intensity of transmitted light. Use this to introduce the **Law of Malus**, $I = I_o cos^2\theta$.

Work several examples (Example 25-5, Active Example 25-2, and others).

➲ DEMO 25-5 Go through the exercise described in Conceptual Checkpoint 25-3 with two crossed polarizers and a third polarizer inserted in the middle. Discuss the result. Repeat using a piece of waxed paper (an "unpolarizer") in the middle instead of the third polarizer sheet.

➲ DEMO 25-6 You can easily do several dramatic examples of **photoelastic stress analysis** with the overhead projector and two crossed polarizers. Again, use the projector as the incandescent light source. Place almost any clear plastic object between the polarizers. (Clear plastic protractors and rulers work quite well, as do clear plastic cups and forks.) As you rotate the polarizers, a brilliant color display shows the places on the object with greatest stress.

Go through the discussion of **liquid crystals**. Point out that liquid-crystal displays are much more energy efficient than LED displays, a critical consideration in battery-operated devices such as calculators. Finish up this section with a discussion of why the sky is blue (see the Real World Physics application) and why polarized sunglasses are better for driving than regular sunglasses. (See Figure 25-22.)

Resource Information

Transparencies

302. Figure 25-1	Producing an electromagnetic wave
303. Figure 25-2	Field directions in an electromagnetic wave
Figure 25-3	The right-hand rule applied to an electromagnetic wave
304. Figure 25-8	The electromagnetic spectrum
305. Figure 25-11	Polarization of electromagnetic waves
306. Figure 25-12	Polarized versus unpolarized light

307.	Figure 25-14	Transmission of polarized light through a polarizer
	Figure 25-15	Transmission of unpolarized light through a polarizer
308.	Figure 25-16	A polarizer and an analyzer
309.	Figure 25-19	Vertically polarized light scattering from a molecule
	Figure 25-20	Unpolarized light scattering from a molecule

Physlet® Illustrations

25-1 Electromagnetic Waves

Suggested Readings

Beaver, J., "The Speed of Light with a Shortwave Radio," *The Physics Teacher* (March 2000), pp. 172-174. Uses shortwave signals from the National Institute of Standards and Technology stations in Colorado and Hawaii to measure the speed of light.

Benenson, R., "Light Polarization Experiments with a Diode Laser Pointer," *The Physics Teacher* (January 2000), pp. 44-46. Introductory polarized light measurements with a diode laser pointer.

Berman, B., "Sky Lights: Bad Day, Sunshine," *Discover* (February 2001), p. 33. A short article on the depletion of the ozone layer and UV levels on Earth.

Brush, S., "Cautious Revolutionaries: Maxwell, Planck, Hubble," *American Journal of Physics* (February 2002), pp. 119-127. Great article on three influential scientists and their attitudes toward their own remarkable discoveries.

Easton, D., "Transmission Through Crossed Polaroid Filters," *The Physics Teacher* (April 2001), pp. 231-233. Describes modifications to digital lux meters for use in Law of Malus and other polarization experiments.

Ellenstein, M., "The Broken-Protractor Prediction Trick," *The Physics Teacher* (January 2002), p. 52. An example of using polarizing sheets to investigate stress lines in molded plastic objects. Includes photographs.

Hosack, H., Marler, N. and MacIsaac, D., "Microwave Mischief and Madness," *The Physics Teacher* (May 2002), pp. 264-266. Lists several short experiments performed by heating objects in a microwave oven.

Kunzig, R., "Trapping Light," *Discover* (April 2001), pp. 72-79. An introductory article on photonics.

Middleton, A. and Sampere, S., "Color Mixing via Polarization," *The Physics Teacher* (February 2001), pp. 123-124. Description of an apparatus for using polarizers to mix colors.

Morris, R., *Dismantling the Universe: The Nature of Scientific Discovery*, Simon and Schuster, New York, 1983. A great book to read before teaching a physics class. Chapter 1, pp. 13-43, covers the story of the ether controversy.

Müller, R., "The Ether Wind and the Global Positioning System," *The Physics Teacher* (April 2000), pp. 243-246. Using GPS to verify the Michelson-Morley experiment.

Ouseph, P., Driver, K. and Conklin, J., "Polarization of Light by Reflection and the Brewster Angle," *American Journal of Physics* (November 2001), pp.1166-1168. Describes an experiment to determine the Brewster angle and to study polarization of reflected light using equipment available from PASCO Scientific.

Perkalskis, B. and Freeman, J., "Demonstrating Induced Birefringence in Stressed Glass," *American Journal of Physics* (September 2000), pp. 871-872. Description of apparatus for demonstrating birefringence.

Prather, E. and Slater, T., "Bringing Extra-Solar Planets into the Introductory Physics Classroom," *The Physics Teacher* (February 2001), pp. 120-122. Describes the use of the Doppler shift for light to discover the existence of planets in other star systems.

Rossing, T. and Chiaverina, C., "Resource Letter TLC-1: Teaching Light and Color," *American Journal of Physics* (October 2000), pp. 881-887. Excellent list of resources for teaching optics.

Materials

Edmund Scientific sells various sizes and types of polarizing sheets. (See, for instance, item CR30384-93.) A polarizer/analyzer kit (item CR30815-70) that can be used with a microscope is also available.

HeNe lasers are available from a variety of sources, including Edmund Optics, model number NT61-315. Many demonstrations can be done with a laser pointer.

Notes and Ideas

*Class time spent on material: Estimated:*______ *Actual:*____________

Related laboratory activities:

Demonstration materials:

Notes for next time:

Chapter 26: Geometrical Optics

Outline

Summary

Geometrical optics concerns the reflection and the refraction of light, both of which cause changes in the direction light is moving. The images formed by various types of mirrors and lenses are examined. Rainbows, mirages, and fiber optics are a few of the applications discussed in this chapter.

Major Concepts

By the end of the chapter, students should understand each of the following and be able to demonstrate their understanding in problem applications as well as in conceptual situations.

- Wave fronts and rays
- Reflection and mirrors
  - The law of reflection
  - Plane mirrors
  - Spherical mirrors – concave and convex
  - Ray tracing and the mirror equation
- Refraction and lenses
  - The law of refraction
  - Total internal reflection
  - Reflection and polarization
  - Thin lenses – converging and diverging
  - Ray tracing and the thin-lens equation
- Dispersion and the rainbow

Teaching Suggestions and Demonstrations

Before beginning with the topics of reflection and refraction, give students a brief overview of optics in general. Distinguish between geometrical optics, the topic of this chapter, and physical optics, which will be covered in Chapter 28. To give students a preview of things to come, mention the wave-particle duality of light and emphasize that the current discussion deals with the wave characteristics of light. Throughout this chapter, keep a collection of lenses and mirrors on hand for quick demonstrations. Glass prisms and a laser are also handy for illustrating refraction and total internal reflection.

Sections 26-1 through 26-4

Figures 26-1 and 26-2 are very helpful in defining **wave fronts** and **rays**. Since parallel rays are used in investigating lenses and mirrors, be sure that students understand that **planar wave fronts** arise when the source is very far away. In geometrical optics, the rays are sufficient for illustrating the phenomena, but it's a good idea to occasionally throw in some wave fronts to remind students what the rays represent.

Include fronts as well as rays the first time you illustrate the **law of reflection**. One reason is that it helps convince students what a convenient simplification the rays provide, since the diagram with fronts is much more cluttered and harder to follow. Fronts also provide an illustration of the physical basis for the law of reflection. With an angle of incidence other than zero, students can see that one "end" of a front hits the surface before the other, causing a change in the direction of the light.

➲ **DEMO 26-1** Shine a laser onto a rectangular or trapezoidal glass prism at varying angles of incidence. The laser beam essentially acts like a ray, and students can compare the **reflected ray** and the **incident ray** at each surface.

We are so used to looking in **plane mirrors**, it doesn't really occur to students to think about the location of the image. A baby or a puppy, though, when confronted with its own image in a mirror, will try to look behind the mirror to find the playmate! Conceptual Checkpoint 26-1, concerning the minimum height of a "full length" mirror, is a fun one to do together. Extend it by asking what other factors determine whether or not you can actually see your whole self. Often, students will think that the distance between the person and the mirror matters. It doesn't, but the vertical placement of the mirror does.

A plane mirror always forms the same type of image. (See the summary of its properties following Example 26-1.) Table 26-1 summarizes the characteristics of the images formed by **convex and concave spherical mirrors**. Point out to students that the image of a convex mirror is always smaller, upright and virtual, but the properties of the image of a concave mirror depend on the object distance as compared to the focal length. Do an example of each case listed in Table 26-1 for the concave mirror using the **mirror equation** and the formula for **magnification**. Then draw **ray diagrams** to scale to check the results for both image distance and image height. Encourage students to practice plenty of ray diagrams on their own, particularly those for virtual images.

➲ **DEMO 26-2** Pass around both **a convex and a concave spherical mirror** and have students examine their own images. For the concave mirror, suggest that they start with the mirror at arm's length and watch the image as they slowly bring it closer and closer. When the object distance equals the focal length, the image will momentarily disappear. Once the object (the student's own face) is inside the focal point, the image will be magnified, upright and virtual.

Students may have trouble with the concepts of **real and virtual images**. If the image can be focused on a screen, then it must be real since the light rays actually pass through it. An image that is made by diverging rays is a virtual image. Light does not actually reach a virtual image, but your brain traces the diverging rays backward to the point from which they appear to have come.

Sections 26-5 through 26-7

Illustrate the **refraction of light** by drawing wave fronts incident on a boundary between two different media. Your picture will look something like the marching band analogy illustrated in Figure 26-20. Then add the rays, which are perpendicular to the fronts. Since the light travels at different speeds in different media, one end of each front will slow down or speed up before the other, causing the path of the light, or ray, to bend. Once students understand the reason for the bending, drop the fronts from the diagrams and work only with the rays. Every once in a while sketch in a few fronts to remind students why the rays behave as they do when entering different media.

Go over the **features of refraction** summarized following Exercise 26-4. Sketch the fronts for incident light traveling along the normal. The fact that the light is unbent follows from Snell's law as well as from the sketch. If light is traveling along a normal to the surface boundary, then all parts of the front change speed at the same time, resulting in a change of wavelength but not of direction.

➲ **DEMO 26-3** Shine a laser beam onto a rectangular or trapezoidal glass prism with the light incident on one of the parallel sides. Now students can examine the **refracted rays** as well as the reflected rays at each surface. As the light travels from air to glass the ray is bent towards the normal, but as it travels from glass to air on the far side of the prism it is bent away from the normal. Point out also that both rays in air are parallel to each other, but the outgoing ray is slightly deviated from a straight-line extension of the incident ray. A triangular glass prism and the laser can be used to demonstrate Conceptual Checkpoint 26-4.

Use both a diagram and Snell's law to help explain why **total internal reflection** cannot occur when light in a medium with a lower index of refraction is incident on a medium with a higher index of refraction. Consider, for example, light traveling from air to glass. Since the ray bends toward the normal, the angle of refraction can never reach 90°. Attempting to solve for the critical angle in this case leaves you trying to find the inverse sine of a number greater than one, which is impossible.

➲ **DEMO 26-4** The trick to demonstrating **total internal reflection** is to use a semicircular prism. Shine the laser in the curved side along a radius. The light will pass straight through that boundary, since the angle of incidence is zero. The angle at which the ray hits the flat side, however, depends upon which radius you choose. Show the path the light follows for angles less than, equal to, and greater than the critical angle. An alternative method consists of placing the prism on a corkboard with a straight pin placed vertically at the center of the flat side of the semicircle. Have students sight through the prism to the pin. As they increase the viewing angle, the pin will suddenly disappear!

➲ **DEMO 26-5** Inexpensive **fiber optic toys** are available through several distributors. (See Resource Information.) Students enjoy seeing these demonstrated. Also mention biological applications, as discussed in the Real World Physics application on optical fibers and endoscopes.

The discussion of **converging and diverging thin lenses** is very similar to the treatment of concave and convex mirrors earlier in the chapter. For convenience, ray diagrams are drawn as though the bending of the light occurs in the center of the lens. Of course it doesn't, because there is no boundary between different media at the center. Be sure to draw some rays refracting at the surfaces as they actually do, so that students will understand why lenses that are thicker in the middle cause parallel light rays to converge while lenses that are thinner in the middle cause parallel rays to diverge.

The equations relating **magnification** and **focal length** to object and image distances are conveniently the same for lenses and mirrors. The **sign conventions** are also the same. The only difference is that the virtual image of a lens is on the same side as the object and a real image is on the opposite side. Light really does pass through a lens, so real objects are formed on the opposite side. Light does not pass through a mirror, so virtual images are on the opposite side. Make sure students understand that the sign of the image distance corresponds to whether or not the image is real or virtual, not the side of the lens or mirror on which it appears. Emphasize the importance of the sign conventions and of familiarity with the types of images formed. For instance, in Example 26-7, the problem concerns "a real image that is twice as large as the object." At first, students may assume that the magnification is positive two. However, real images are inverted so the correct magnification is negative two.

➲ DEMO **26-6** Use various **converging lenses** to project the image of a light source onto a screen. Begin with a very large object distance, which will result in a tiny image that is formed very close to the focal point. Have students watch what happens to the image as the distance between the light and the lens decreases. Once the light is at a distance less than the focal length, it will be impossible to project the image onto a screen.

➲ DEMO **26-7** Students can also look through both **converging and diverging lenses** and describe and compare the images they see for different object distances. A converging lens used with an object inside the focal point is simply a magnifying glass, which is discussed further in Chapter 27. The resulting image is upright, magnified and virtual.

➲ DEMO **26-8** An optics set that includes a ray projector and various lenses and mirrors is an excellent tool for demonstrating **geometrical optics**. Many sets come with magnets or suction cups that allow you to attach the elements to the board. You can use the set to demonstrate the laws of reflection and refraction and the behavior of light rays as they interact with mirrors and lenses. (See Resource Information.)

➲ DEMO **26-9** Students often mistakenly believe that research-grade astronomical telescopes are so large because magnification depends on the **lens diameter**. Find two converging lenses with the same focal length but different diameters. Dim the room lights and use each to project an image of a window and the scene outside it on the opposite wall. First, the students will notice that all the trees and cars framed by the window, and indeed, the window frame itself, are upside down. Now have students compare the two different images. The magnification will be the same, but the image formed by the larger diameter lens will be brighter and thus show more detail.

Section 26-8

"Blue bends best; red refracts rotten" is a handy saying to help students remember how the index of refraction for a given material depends on the color of the light. Students will be familiar with the **dispersion** caused by a triangular prism. Draw a diagram of white light incident on a slab of glass with parallel faces and show them that in this case, too, the colors are dispersed. However, the rays representing the different colors all leave the glass parallel to one another and are so slightly displaced that we can't distinguish the colors. The reason a prism produces a rainbow is that the rays are diverging when they leave the glass.

Resource Information

Transparencies

310. Figure 26-2 Spherical and planar wave fronts

311.	Figure 26-4	Reflection from smooth and rough surfaces
312.	Figure 26-6	Locating a mirror image
313.	Figure 26-10	Parallel rays on a convex mirror
	Figure 26-11	Ray diagram for a convex mirror
314.	Figure 26-12	Parallel rays on a concave mirror
315.	Figure 26-14	Principal rays used in ray tracing for a concave mirror
	Figure 26-15	Principal rays used in ray tracing for a convex mirror
316.	Figure 26-16	Image formation with a convex mirror
317.	Figure 26-18	Image formation with a concave mirror
318.	Figure 26-19	Ray diagrams used to derive the mirror equation
319.	Figure 26-20	An analogy for refraction
	Figure 26-21	The basic mechanism of refraction
320.	Exercise 26-4	
321.	Figure 26-24	Light propagating through a glass slab
322.	Figure 26-25	Total internal reflection
323.	Figure 26-29	A variety of converging and diverging lenses
324.	Figure 26-34	The image formed by a concave lens
325.	Figure 26-35	Ray tracing for a convex lens
326.	Figure 26-36	Ray diagrams used to derive the thin-lens equation
327.	Figure 26-37	Dispersion in a raindrop
	Figure 26-38	How rainbows are produced

Physlet® Illustrations

26-1 Law of Reflection
26-2 Plane Mirror
26-3 Convex Mirror
26-4 Concave Mirror
26-5 Law of Refraction
26-6 Apparent Depth
26-7 Convex Lens
26-8 Dispersion in a Prism

Suggested Readings

Chakravarti, S. and Siegel, P., "Visualizing the Thin-Lens Formula," *The Physics Teacher* (September 2001), pp. 342-343. Uses a graph of $1/i$ versus $1/o$ to help students investigate and verify the thin-lens formula.

Gatland, I., "Thin Lens Ray Tracing," *American Journal of Physics* (December 2002), pp. 1184–1186. Presents a ray-tracing approach to thin lens analysis.

Greenslade, T., "A Quick Experiment on Reflection from Concave Mirrors," *The Physics Teacher* (April 2000), p. 206. Describes how to make inexpensive concave mirrors and use them for simple experiments.

Helfgott, H. and Helfgott, M., "A Noncalculus Proof That Fermat's Principle of Least Time Implies the Law of Refraction," *American Journal of Physics* (December 2002), pp. 1124–1125. An algebraic proof that shows how Fermat's principle of least time implies Snell's law.

Houser, B., "Demonstrating the Decreased Wavelength of Light in Water," *The Physics Teacher* (April 2001), pp. 228-229. A good demonstration of what is and is not constant about light as it moves from one medium to another.

Johnson, A. and Hitz, C., "Career Opportunities in Optics," *Physics Today* (May 2000), pp. 25-28. An interesting article about advances and applications in the area of optical communications and the resulting increase in career possibilities.

Kakaes, K., "Theorists Doubt Claim for Perfect Lens," *Science* (24 May 2002), v. 296, pp. 1380-1381. A short, interesting article on materials with negative indices of refraction.

Lawrence, R., "Magnification Ratio and the Lens Equations," *The Physics Teacher* (March 2000), pp. 170-171. Presents a method of quickly visualizing relationships of object and image distances to magnification.

Newburgh, R., Rueckner, W., Peidle, J. and Goodale, D., "Using the Small-Angle Approximation to Measure the Index of Refraction of Water," *The Physics Teacher* (November 2000), pp. 478-479. A nice variation of a standard introductory physics lab.

O'Connell, J., "Optics Experiments Using a Laser Pointer," *The Physics Teacher* (October 1999), pp. 445-446. Nine simple demonstrations of geometrical and physical optics.

Sawicki, M. and Sawicki, P., "Supernumerary Rainbows," *The Physics Teacher* (January 2000), p. 19. Beautiful photograph of multiple rainbows.

Shelby, R., Smith, S. and Schultz, S., "Experimental Verification of a Negative Index of Refraction," *Science* (6 April 2001), v. 292, pp. 77-79. An advanced article on a material with a negative index of refraction.

Tape, C., "Aquarium, Computer, and Alaska Range Mirages," *The Physics Teacher* (May 2000), pp. 308-311. Photographs and descriptions of mirages.

Thompson, G. and Mathieson, D., "The Mirror Box," *The Physics Teacher* (November 2001), pp. 508-509. Detailed instructions for constructing a mirror box suitable for class demonstrations of the law of reflection and for magic tricks.

Materials

The "Blackboard Optics™ Basic Set" from Klinger (item KO4100 or KO4100M) is one example of an assembly of lenses, mirrors and ray projectors that can attach right to your blackboard. An additional accessory set (KO4100A or KO4100AM) provides components for more demonstrations and experiments in optics.

Edmund Optics can be found online at www.edmundoptics.com. It is a great source for lenses, prisms, and other optical components.

Fun fiber-optic demonstrations include “lumirods” from Fisher Science Education (item SA1678-1, for example) and fiber-optic light wands (item CR30317-71) from Edmund Scientific.

Notes and Ideas

Class time spent on material: Estimated:______ Actual:______________

Related laboratory activities:

Demonstration materials:

Notes for next time:

Chapter 27: Optical Instruments

Outline

Summary

The human eye is a remarkable optical instrument. Chapter 27 applies the geometrical optics from Chapter 26 to explain how the eye creates an image. Optical instruments that correct or extend our vision, such as eyeglasses, contact lenses, magnifying glasses, microscopes, and telescopes, are explored. Finally, lens aberrations and ways to correct for them are discussed.

Major Concepts

By the end of the chapter, students should understand each of the following and be able to demonstrate their understanding in problem applications as well as in conceptual situations.

- The human eye and the camera
  - Focusing and image formation
  - The *f*-number
  - Nearsightedness
  - Farsightedness
- The magnifying glass
- Instruments with lens combinations
  - The compound microscope
  - The telescope
- Lens aberrations and corrections

Teaching Suggestions and Demonstrations

You very likely will not have much time to spend on this chapter, but try to at least hit the highlights. Students, especially those with an interest in biology and/or those who wear glasses, typically find the study of the human eye fascinating. Likewise, cameras, telescopes and microscopes provide a wonderful opportunity to apply geometrical optics to practical instruments with which the students are familiar.

Sections 27-1 and 27-2

The **parts of the eye**, **image formation**, and **accommodation** are illustrated in Figures 27-1 through 27-3. Point out the difference in focusing between an eye and a **camera**. The muscles in the eye change the shape of the lens and thus its focal length. The focal length of a camera lens is fixed, so focusing is accomplished by moving the entire lens toward or away from the film.

➲ **DEMO 27-1** Show students a simple **camera** and have them compare it to an eye. In addition to the focusing differences mentioned above, compare the camera's adjustable aperture to a person's pupil, and the film to the retina.

➲ **DEMO 27-2** The **pinhole camera**, or *camera obscura*, remarkably forms an image without a lens. Make a small viewer by forming a rectangular box without a front or back out of black poster board. For the front, cut another piece of black poster board and put a pinhole in the center. (Alternatively, cut a small square out of the front piece, cover it with aluminum foil and put the pinhole in the foil.) Cover the back with waxed paper. (This is the view screen.) Seal all the edges with black electrical tape. Stand in a darkened room and aim the viewer at the window. An image of the window scene will form on the waxed paper. Keep the box small so that the image is bright enough to see. Sketch a ray diagram for students to clarify why an image forms and also why the hole is necessarily small.

For safe solar-eclipse viewing, poke a small hole in a piece of cardboard and project an image of the sun onto a sheet of white paper. (Remind students to never look directly at the sun!)

Since the rest of the chapter deals with **combinations of lenses**, do Active Example 27-2 or one like it thoroughly. An important idea to get across to the students is that the image of the first lens acts as the object for the second lens. Demonstrate finding the image by both the thin-lens equation and by using a ray diagram.

A "normal" human eye can focus on any object between about 25 cm away (the near point) and infinity (the far point). That's quite an impressive range! Of course, many people don't have this range and therefore wear glasses or contact lenses to correct their **nearsightedness** or **farsightedness**. In either case, the external lens and the eye work together as two lenses in combination. The external lens forms an image between the near and far points, and then the eye uses that image as its object. The **refractive power** of a lens can be confusing. Point out that it is just the inverse of the focal length and therefore its units, diopters, are simply inverse meters. If you are short on time, skip the quantitative discussion of corrective lenses and just illustrate with diagrams such as those in Section 27-2.

➲ **DEMO 27-3** There are no rods or cones at the location of the optic nerve, so if an image happens to fall there, no message gets sent to the brain and you can't see the object in question. To demonstrate this **blind spot**, have students do the following. Mark two dark X's 10 to 15 cm apart on a piece of paper. With the paper stretched out in front of you at arm's length, close your left eye and look with your right eye at the left-hand X. (The right-hand X will be visible farther to the right with your peripheral vision.) Now slowly bring the paper toward your face, focusing on the left-hand X but noticing the right-hand X out of the corner of your eye. The right-hand X disappears at the point when its image falls on your blind spot.

Sections 27-3 through 27-5

A **magnifying glass** is a simple convex lens like those discussed in Chapter 26. It works essentially by allowing the viewer to move the object closer to the eye but still focus it. Discuss the definition of **angular magnification** and compare it to the definition of magnification given in Chapter 26.

Both **compound microscopes** and **refracting telescopes** consist of an objective lens and an eyepiece. The eyepiece is a magnifier, which uses the image of the objective as its object. Go over the arrangements of the lenses as shown in Figures 27-15 and 27-16 and compare the two instruments, pointing out the difference in object distance. Again, if you are short on time, concentrate on the conceptual diagrams and skip the quantitative analysis.

➲ DEMO 27-4 **Telescopes** can be demonstrated quite easily with a couple of lenses, a meter stick and some clay. Place the lenses on the horizontal meter stick and hold them in place vertically with the clay. Example 27-6 will help students determine how far apart to put the lenses. Then have students hold the telescope up and sight through the pair of lenses at an object across the room or outside. It takes some practice but they should be able to view magnified images. The magnification is hard to estimate; counting bricks on a brick wall or examining some other regular pattern helps. Repeat using lenses with different focal lengths and compare. Also try a diverging lens as the eyepiece.

A good example of a lens-mirror combination is the **reflecting telescope**, which has a mirror as its objective and a lens as its eyepiece. Figure 27-17 shows one possible arrangement of the two optical elements. For telescopes, a larger objective is better since most objects of astronomical interest are dim and larger apertures collect more light. As discussed at the end of Section 27-5, large mirrors are easier to make than large lenses, so the best astronomical research telescopes in the world are reflectors. An ingenious method of making telescope mirrors, called "spin casting," involves melting glass in a huge, spinning oven. The rotation helps form the curvature of the mirror.

Section 27-6

The two main types of **lens aberrations**, chromatic and spherical, can be covered very briefly. Students may have observed chromatic aberration while looking through a telescope eyepiece. Note that chromatic aberration is not a problem with mirrors, since the angle of reflection is not wavelength dependent.

Resource Information

Transparencies

328. Figure 27-1	Basic elements of the human eye
329. Figure 27-3	Accommodation in the human eye
330. Figure 27-4	Basic elements of a camera
331. Figure 27-5	A two-lens system
332. Figure 27-6	Eye shape and nearsightedness
Figure 27-7	Correcting nearsightedness
333. Figure 27-10	Eye shape and farsightedness
Figure 27-11	Correcting farsightedness
334. Figure 27-13	How a simple magnifier works
335. Figure 27-15	The operation of a compound microscope
336. Figure 27-16	Basic elements of a telescope
337. Figure 27-18	Spherical aberration
Figure 27-19	Chromatic aberration in a converging lens

Physlet® Illustrations

27-1 Nearsightedness
27-2 Farsightedness
27-3 Microscope

Suggested Readings

Andereck, B. and Secrest, S., "The Magic Magnifier," *The Physics Teacher* (May 2001), pp. 301-302. Describes the construction of a simple telescope out of PVC pipe and lenses.

Graf, E., "How Do *You* Use a Magnifying Glass?," *The Physics Teacher* (May 2001), pp. 298-300. Compares standard explanations of the use of magnifying glasses to the way people really use them.

Henderson, H., "A Dialog in Paradise: John Milton's Visit with Galileo," *The Physics Teacher* (March 2001), pp. 179-183. An essay about Milton, his literary works, and his visit with Galileo.

Huebner, J., Gibbs, D. and Ryan, P., "Projecting Chromatic Aberrations," *American Journal of Physics* (September 2000), pp. 869-870. Describes apparatus for demonstrating chromatic aberration to a large class using an overhead projector.

Knight, J. and Russell, P., "New Ways to Guide Light," *Science* (12 April 2002), v. 296, pp. 276-277. A good short article on new optical fibers.

McLellan, J., Marcos, S., Prieto, P. and Burns, S., "Imperfect Optics May Be the Eye's Defense Against Chromatic Blur," *Nature* (9 May 2002), v. 417, pp. 174-176. An advanced but accessible treatment of the optics of the eye.

Miller, D., "Retinal Imaging and Vision at the Frontiers of Adaptive Optics," *Physics Today* (January 2000), pp. 31-36. Good discussion of the potential for adaptive optics to correct for aberrations, diffraction and the eye's defects.

Milonni, P., "Resource Letter: AOA-1: Adaptive Optics for Astronomy," *American Journal of Physics* (June 1999), pp. 476-485. A guide to the basic concepts and the literature on adaptive optics for astronomy.

Porter, N., *Physicists in Conflict*, Institute of Physics Publishing, Bristol and Philadelphia, 1998. Chapters 3 and 4 discuss Galileo and Kepler.

Salinas, J. and Sandoval, J., "Geometrical Optics and Visual Perception," *The Physics Teacher* (October 2001), pp. 420-423. An examination of the relationship between images perceived by an observer and the curvature of the light wave fronts reaching the eye.

Thomas, G., Ackerman, D., Prucnal, P. and Cooper, S., "Physics in the Whirlwind of Optical Communications," *Physics Today* (September 2000), pp. 30-36. Basics of digital optical communications as well as descriptions of new innovations.

Wagner, D. and Walkiewicz, T., "When the Eye Meets the Lens," *The Physics Teacher* (November 2000), pp. 474-475. Describes experiments to clarify the confusion between the observer-independent magnification of a lens and the magnifying power as seen by a particular observer.

Wardle, D., "The Time Delay in Human Vision," *The Physics Teacher* (October 1998), pp. 442-444. A nice demonstration of reaction time and human vision.

Notes and Ideas

Class time spent on material: Estimated:______ Actual:______________

Related laboratory activities:

•

Demonstration materials:

Notes for next time:

Chapter 28: Physical Optics: Interference and Diffraction

Outline

Summary

Chapter 28 discusses two major properties of light that illustrate its wave characteristics, interference and diffraction. Superposition of waves was first discussed in Chapter 14 and is applied to electromagnetic waves in this chapter. Double-slit interference, single-slit diffraction, and thin-film interference are the main topics covered.

Major Concepts

By the end of the chapter, students should understand each of the following and be able to demonstrate their understanding in problem applications as well as in conceptual situations.

- Superposition
- Interference
  - Young's two-slit experiment
  - Air wedge
  - Newton's rings
  - Thin-film interference
- Diffraction
  - Single-slit diffraction
  - Resolution
  - Diffraction gratings

Teaching Suggestions and Demonstrations

Students won't necessarily be able to visualize interference patterns until they see them, so bring out the laser and do lots of short demonstrations of double-, single-, and multiple-slit interference throughout the chapter. Continually remind students that they are seeing wave phenomena. Experiments discussed in Chapter 30 will illustrate the particle nature of light. Students will probably ask how light can be both a particle and a wave, and how it knows when to act like a particle and when to act like a wave. Begin these discussions of wave-particle duality now and continue them in the chapters on modern physics.

Sections 28-1 and 28-2

Students should already be familiar with **wave superposition and interference** from their study of waves on a string and sound waves in Chapter 14. Take a little time at the beginning of this chapter to review these concepts and then apply them to electromagnetic waves. The ideas of **phase** and **phase difference** are critical to the understanding of the interference of light waves, so be sure students have a firm grasp of these terms.

The conditions necessary for interference effects of light to be noticeable are also important to discuss. Most students understand the idea of **monochromatic** light fairly easily. However, the concept of **coherent light sources** is a little trickier. Point out that the two sources don't necessarily need to be in phase, but they do need to maintain a constant phase difference. Overlapping light waves will interfere no matter what, but if the sources are incoherent the interference will vary rapidly and randomly so that no overall interference pattern will be noticeable.

Example 28-1 is essentially a **Young's two-slit experiment** done with radio waves emitted by two antennae instead of light emerging from two slits. Students will find Young's experiment with light easier to understand if you go over the radio wave example first. Also refer back to Figure 14-21 to illustrate the interference pattern of two overlapping circular waves.

Figure 28-3 shows the setup for Young's experiment. Point out the purpose of the initial slit. Young performed his experiment before lasers were available to produce coherent light. Next, ask students what the particle model of light would predict from the experiment. If light were a stream of tiny particles, you would expect them to pile up on the screen directly across from each slit instead of resulting in the alternating bands of dark and light. Go over the derivations for the **conditions for bright and dark fringes** carefully. In particular, make sure students understand the concept of **path difference** and that the relationship between the path difference and the wavelength is what determines whether constructive or destructive interference occurs.

➲ **DEMO 28-1** A set of slides for **demonstrating interference and diffraction** is well worth purchasing. (See Resource Information.) Sets usually include single slits, pairs of slits, and multiple slits as well as circular apertures. Shine a laser through a double slit to project the interference pattern on a wall. Point out to students that you don't need the initial slit as in Young's original experiment because your source is already coherent. In actuality, the pattern will be a combination of the double-slit pattern with a single-slit pattern superimposed, because the two slits are each of finite width. You can mention this and return to the demonstration when you discuss single-slit diffraction later. After students have examined the pattern, ask them to predict what will happen if you use a smaller slit separation. Using other slides from the set, demonstrate both smaller and larger slit separations.

Section 28-3

By now students will understand the importance of path difference to the interference of two light waves. For the **interference of reflected waves**, emphasize that a second factor must also be taken into account—namely, the possible **phase change due to reflection**. The combination of these two factors will determine if the interference is constructive or destructive. Refer back to Chapter 14, and particularly Figure 14-20, for the mechanical analogy of phase change due to reflection.

Thin-film interference can be confusing for students because the condition for destructive interference in some cases is the condition for constructive interference in others. The key is whether a phase shift on reflection occurs at only one or at both interfaces, which depends on the relative indices of refraction. To clarify the difference, first derive the interference conditions for films of negligible thickness. In this case, only phase shifts matter. Then add thickness to the film and take into account the path difference as well, with which students are already familiar from Young's two-slit experiment. Point out that when comparing path difference to wavelength, the wavelength *in the medium*, not in air, is what matters. The colors in soap bubbles and oil slicks are two very common examples of thin-film interference. In addition, the interference of reflected waves can be used for practical applications, as discussed for Newton's rings, air wedges, and CD players.

➲ DEMO 28-2 Students certainly will have seen the **colors of soap bubbles** before, but it's fun to pass out a few jars of bubble liquid and blowers and have them take a closer look. As the thickness of a bubble changes from point to point, so do the colors. If the humidity is right and the bubbles last a while before popping, you can even see what happens as the soap film becomes negligibly thin at the top of the bubble.

➲ DEMO 28-3 Get a brand-new package of microscope slides and separate them in pairs. Tape each pair together near the short edges. A very thin film of air will be trapped between the slides. If you press down in the center with a dull pencil point, you will see **Newton's rings** surrounding the point of pressure.

Sections 28-4 through 28-6

Have students compare Figure 28-17, showing the **diffraction** of water waves, to Figure 28-4, showing **Huygen's principle**. Light diffracts as it passes through the narrow slits. Emphasize that diffraction is a wave phenomenon and thus illustrates the wave nature of light. The idea of wave-particle duality bears repeating, since it is completely non-intuitive.

It makes sense to students that light from two different slits will interfere, but the idea that light passing through a **single slit** will interfere with itself to produce dark and light bands is more difficult. Examine Figures 28-19 and 28-20 with students, and use them to derive Equation 28-12, the condition for a dark fringe in single-slit interference.

➲ DEMO 28-4 Shine a laser through a **single slit** and examine the pattern produced. Have students try Conceptual Checkpoint 28-3, regarding the width of the slit and the width of the central maximum. The result that a narrower slit produces a wider central maximum is important and not necessarily obvious. Compare patterns formed by single slits of different widths. Then return to the double-slit interference pattern and show students the effect of the individual slit widths on the overall pattern.

➲ DEMO 28-5 To introduce **resolution** and **Rayleigh's criterion**, shine a laser through a circular aperture and note the resulting circular diffraction pattern, as shown in Figure 28-21. In addition, try making a poster of pairs of dots with different separations. Have students start far away from the poster and see which pairs they can resolve. As they move closer, they will be able to resolve the more closely spaced pairs.

Extend the discussion of single and double slits to **multiple slits** and you have a diffraction grating. Point out the advantages of diffraction gratings over prisms for spreading light into its colors.

➲ DEMO 28-6 Use the laser and a slide with multiple slits to demonstrate **diffraction gratings**. Have students predict and then examine what happens as the number of slits and the slit spacing change.

➲ DEMO 28-7 **Diffraction grating glasses** are inexpensive to buy and lots of fun. (See Resource Information.) Students put on the glasses and view different light sources. Have them look at extended sources, such as the lights in the classroom, as well as narrow sources. In Chapter 31, students can use the glasses again to view spectral lines of different gases.

Resource Information

Transparencies

338. Figure 28-1	Constructive and destructive interference
339. Figure 28-3	Young's two-slit experiment
340. Figure 28-4	Huygen's principle
Figure 28-5	Path difference in the two-slit experiment
341. Figure 28-6	The two-slit pattern
342. Figure 28-8	Phase change with reflection
343. Figure 28-9	An air wedge
344. Figure 28-12	Interference in thin films
345. Figure 28-14	A thin film with one phase change
Figure 28-15	A thin film with two phase changes
346. Figure 28-18	Single-slit diffraction
347. Figure 28-19	Locating the first dark fringe in single-slit diffraction
348. Figure 28-22	Resolving two point sources: Rayleigh's criterion

Physlet® Illustrations

28-1 Two-Slit Interference
28-2 Thin-Film Interference
28-3 Polarization

Suggested Readings

Ambrose, B., Heron P., Vokos, S. and McDermott, L., "Student Understanding of Light as an Electromagnetic Wave: Relating the Formalism to Physical Phenomena," *American Journal of Physics* (October 1999), pp. 891-898. Development and modification of tutorials to address student difficulty with the wave nature of light.

Ambrose, B., Shaffer P., Steinberg, R. and McDermott, L., "An Investigation of Student Understanding of Single-Slit Diffraction and Double-Slit Interference," *American Journal of Physics* (February 1999), pp. 146-155. Discusses evidence of student misunderstanding of both the wave nature of light and the photon, with implications for curriculum development.

Gallis, M., "Automating Microwave Optics Experiments," *The Physics Teacher* (April 2002), pp. 217-219. Describes two types of microwave experiments that demonstrate interference phenomena.

Hernández-Andrés, J., Valero, E., Nieves, J., and Romero, J., "Fizeau Fringes At Home," *American Journal of Physics* (July 2002), pp. 684–688. A description of an easy, inexpensive method to obtain Fizeau fringes using a flatbed scanner and a personal computer.

Kovács, A., Varjú, K., Osvay, K. and Bor, Z., "On the Formation of White-Light Interference Fringes," *American Journal of Physics* (November 1998), pp. 985-989. An experiment to investigate white-light interference fringes using a Michelson interferometer and a spectrograph.

Moloney, M., "Homemade Interference," *The Physics Teacher* (November 1999), pp. 504-505. Uses the interference of two tiny light sources viewed through a cloth grating to determine the wavelength of light.

Perkalskis, B. and Freeman, R., "Herschel's Interference Demonstration," *The Physics Teacher* (March 2000), p. 142. Apparatus for using two prisms to demonstrate multiple interference.

Rueckner, W. and Papaliolios, C., "How to Beat the Rayleigh Resolution Limit: A Lecture Demonstration," *American Journal of Physics* (June 2002), pp. 587-594. Describes an experiment to demonstrate the effect of instrument aperture size on resolution, as well as the use of interferometry to detect the presence of more than one object even when they cannot be resolved by the instrument. Includes photographs.

Sawicki, C., "Easy and Inexpensive Demonstration of Light Interference," *The Physics Teacher* (January 2001), pp. 16-19. Uses interference fringes to measure the thickness of plastic wrap.

Wein, G., "A Video Technique for the Quantitative Analysis of the Poisson Spot and other Diffraction Patterns," *American Journal of Physics* (March 1999), pp. 236-240. Presentation of an approach for the observation and analysis of diffraction phenomena using a video camera.

Materials

The "Introductory Optics System" from Pasco (item OS-8500) is a complete set including equipment for demonstrating geometrical optics as well as interference and diffraction.

The "Diffraction Optics Kit" (Pasco item OS-8531) includes only those components necessary for demonstrating diffraction.

The "Slide Set and Gratings" available from Fisher Science Education (item S42539) contains six different slides with various numbers of slits and slit separations.

The "Deluxe Laser Accessories Kit" (Fisher item S42701ND) is a much more comprehensive set, including circular apertures, a prism, various slits and diffraction gratings, a hologram, other accessories, and a book detailing 30 experiments.

Edmund Scientific has diffraction-grating film available for purchase by the roll (item CR30521-16).

Diffraction-grating glasses can be ordered from Fisher Science Education. They are listed as diffraction viewing glasses, catalog number NCS48814 for individual glasses, or catalog number NCS48814A for a package of 10.

Notes and Ideas

*Class time spent on material: Estimated:*______ *Actual:*______________

Related laboratory activities:

Demonstration materials:

Notes for next time:

Chapter 29: Relativity

Outline

Summary

Chapter 29 begins the study of "modern" physics. Relativity governs the behavior of systems moving at large velocities, and quantum mechanics, the subject of Chapter 30, deals with microscopic systems. All of these new laws must reduce to the more familiar Newtonian mechanics as the systems enlarge to macroscopic size and slow down to everyday speeds. The postulates of special relativity are presented, and their consequences, including time dilation, length contraction and mass increase, are derived. Einstein's famous equation $E = mc^2$ is also discussed. The chapter ends with a presentation of general relativity, which applies to accelerating reference frames and gravitation.

Major Concepts

By the end of the chapter, students should understand each of the following and be able to demonstrate their understanding in problem applications as well as in conceptual situations.

- Special relativity
  - The two postulates
  - Time dilation
  - Length contraction
  - Mass increase
  - Relativistic addition of velocities
- Mass-energy equivalence and $E = mc^2$
  - Matter and anti-matter
  - Relativistic kinetic energy
- General Relativity
  - The equivalence principle

Teaching Suggestions and Demonstrations

Relativity is an incredibly fascinating (and difficult) subject to teach. Students will most likely not believe you at first, and exchange "let's just humor the professor" looks as you attempt to convince them of time dilation and length contraction. Be sure to include details of experimental evidence verifying relativity throughout your presentation of this chapter. You can begin by making some general points regarding relativity to set the stage and address misconceptions students may have. Relativity is about the nature of space and time. Relativity is not difficult mathematically, but it is very difficult conceptually because we do not have any day-to-day experiences with macroscopic objects traveling near the speed of light. Relativity is not an optical illusion created by high-speed travel.

Section 29-1

As an overview of the next few sections, tell students that you will present the two **postulates of special relativity**, see what consequences result from them, and then look for experimental verification of those consequences. The idea is not to prove the postulates, since postulates are by definition statements accepted without proof, but to accept them and then see where they lead. Any "proof" of relativity will come as experiments are compared to predictions. You will need to spend some time going over the postulates. For the first, emphasize that if you are in an inertial reference frame, there is no experiment you can perform to show if you are moving or at rest. For the second, point out that nothing can travel faster than *c*.

Give the students a hint of what's to come by examining the postulates a little further. Notice that the speed of, say, a baseball depends on the reference frame of the observer. The batter will see a different speed than someone racing by the ball field in a fast car. However, this does not apply to light; the speed of light is the same for all observers. Distance equals speed x time, so if the speed stays constant, weird things must happen to distance and time.

Sections 29-2 through 29-5

Introduce the consequences that follow from the postulates of special relativity by deriving Equation 29-2 for **time dilation** carefully, using Figures 29-5 and 29-6. Point out explicitly when you are using the postulates. Note that the math isn't difficult, but the results are hard to believe! Do a number of examples to help students recognize which time is the proper time in each situation. Point out that "proper" does not mean "correct." Students will be tempted to think that the observer in the rest frame of the event is right and the other observer is wrong, which is not the case; both are right. Another common misconception arising from the phrase "moving clocks run slow" is that somehow the mechanisms in clocks don't work right when they are speeding along. There is nothing wrong with the clocks; time itself is dilated.

Do a problem involving hypothetical high-speed space travel such as Example 29-1 and discuss the results. Be sure to define the **light-year** first and take a moment to calculate just how far a light-year is in kilometers. Of course, we are not capable of accelerating spaceships to anywhere near the speed of light, so it is important to do a realistic example regarding muons or other elementary particles as well. Discuss actual experiments, such as those involving muons performed at CERN.

After deriving the expression for **length contraction,** return to Example 29-1 and re-do it with Equation 29-3. Find the distance traveled according to the people on the spaceship and use it to determine the time of travel. The end result will be the same. Emphasize that in this problem one observer measured the proper time (and a contracted length), while the other observer measured a proper length (and a dilated time). Conceptual Checkpoint 29-2 helps ensure that students understand the direction dependence of the length contraction.

Figures 29-12, 29-13, and 29-16 show the graphs for **relativistic velocity addition, momentum** and **kinetic energy** compared to their classical counterparts. An important point to make about all three of these graphs is that the classical and relativistic curves are essentially the same in each case at lower speeds and diverge as speed increases. Remind students also that "low speed" means low with respect to *c*. Even one-tenth the speed of light, though, is incredibly fast in everyday terms, so we don't notice relativistic effects in our everyday lives. Students sometimes think that Newton's laws apply to slow speeds and Einstein's relativity to fast speeds, and then wonder where to draw the line. Emphasize that the relativistic equations apply to all speeds, but when $v << c$, they reduce to the equations of classical physics.

Discuss the difference between **relativistic mass** and **rest mass** and how the increase of mass with speed is consistent with the previous result that nothing can be accelerated to a speed greater than the speed of light. The result is particularly important in particle accelerators.

Section 29-6

Even students who have never had any physics will come to your class having heard Einstein's famous equation, $E = mc^2$. Do a few quantitative examples to emphasize just how big the number c^2 is. Mention the difference between fusion and fission, which will be covered in more detail in Chapter 32. Many students may think that **antimatter** is an invention of science fiction, so spend a little time discussing it. The positron appears again in the nuclear decay equations of Chapter 32. The PET scanner discussed in this section is a practical application of electron-positron annihilation that will be of particular interest to students in the pre-health professions.

Sections 29-7 and 29-8

General relativity, which applies to accelerated frames of reference and to gravitation, can be covered briefly. Discuss the **principle of equivalence**, using the elevator examples as shown in Figures 29-17 through 29-21. Again, be sure to give experimental verification of relativity, such as Eddington's experiment measuring the displacement of stars during a solar eclipse.

Resource Information

Transparencies

349. Figure 29-2	Wave speed versus source speed
350. Figure 29-3	Wave speed versus observer speed
351. Figure 29-5	A stationary light clock
352. Figure 29-6	A moving light clock
353. Figure 29-7	Time dilation as a function of speed
354. Figure 29-8	A relativistic trip to Vega
355. Figure 29-9	Length contraction
356. Figure 29-12	Relativistic velocity addition
357. Figure 29-13	Relativistic momentum
358. Figure 29-16	Relativistic kinetic energy
359. Figure 29-17	A frame of reference in a gravitational field
Figure 29-18	An inertial frame of reference with no gravitational field
Figure 29-19	An accelerated frame of reference
360. Figure 29-20	A light experiment in two different frames of reference
Figure 29-21	The principle of equivalence
361. Figure 29-22	Gravitational bending of light
362. Figure 29-25	Warped space and black holes
363. Figure 29-26	Gravity waves
364. Figure 29-27	How LIGO detects a gravity wave

Physlet® Illustrations

29-1 Time Dilation
29-2 Length Contraction

Suggested Readings

Ashby, N., "Relativity and the Global Positioning System," *Physics Today* (May 2002), pp. 41-47. Explains the importance of taking fundamental relativistic principles into account for the precise operation of the GPS.

Berlinski, D., "Einstein and Gödel," *Discover* (March 2002), pp. 38-42. An excellent historical article on the friendship between these two men.

Berman, B., "Sky Lights: Curved Space," *Discover* (August 2001), p. 28. A good short article on the bending of light by matter.

Berman, B., "Sky Lights: The Hole Story," *Discover* (October 2001), p. 36. A short article on black holes.

Blackman, E., "Astrophysical Perspectives in Teaching Special Relativity," *The Physics Teacher* (March 1998), p. 177. A nice astrophysical example for special relativity.

Brown, H., "The Origins of Length Contraction," *American Journal of Physics* (October 2001), pp.1044-1054. An analysis of the null result of the Michelson–Morley experiment and its interpretation by both Lorentz and FitzGerald.

Carlip, S., "Kinetic Energy and the Equivalence Principle," *American Journal of Physics* (May 1998), pp. 409-413. A reanalysis of experimental data to test the equivalence principle for the kinetic energy of atomic electrons.

Field, J., "Space–Time Exchange Invariance: Special Relativity as a Symmetry Principle," *American Journal of Physics* (May 2001), pp. 569-575. An advanced article on relativity, with a useful introduction.

Folger, T., "Does the Universe Exist If We're Not Looking?" *Discover* (June 2002), pp. 44-49. Reflections from John Wheeler on observation and experiment.

Greenslade, T., "Relativistic Meter Sticks," *The Physics Teacher* (May 2000), p. 315. A visual display of length contraction.

Gülmez, E., "Measuring the Speed of Light with a Fiber Optic Kit: An Undergraduate Experiment," *The American Journal of Physics* (July 1997), pp. 614-618. A description of an experiment using a commercially available kit in which time-delay is measured to determine the speed of light.

Hecht, E., "From the Postulates of Relativity to the Law of Inertia," *The Physics Teacher* (November 2000), pp. 497-498. Musings on the connections between the postulates of relativity and the law of inertia.

Irion, R., "B-Meson Factories Make a 'Number From Hell'," *Science* (23 February 2001), v. 291, p. 1471. A short article on the potential imbalance of matter and antimatter in the universe.

Kunzig, R., "Black Holes Spin?" *Discover* (July 2002), pp. 32-39. A good general article on new information on black holes.

Lamoreaux, S., "Testing Times in Space," *Nature* (25 April 2002), v. 416, pp. 803-804. An accessible summary of a *Physical Review* article on space-based testing of the constancy of universal constants.

Naddy, C., Dudley, S. and Haaland, R., "Projectile Motion in Special Relativity," *The Physics Teacher* (January 2000), pp. 27-29. Revisiting projectile motion using special relativity considerations.

Osborne, I., Rowan, L., and Coontz, R., " Spacetime, Warped Branes, and Hidden Dimensions," *Science* (24 May 2002), v. 296, p. 1417. A short introduction to a collection of articles on spacetime.

Scheider, W., *Maxwell's Conundrum: A Serious but Not Ponderous Book About Relativity*, Cavendish Press, Michigan, 2000. An algebraic introduction to relativity.

Seife, C., "Crystal Stops Light in Its Tracks," *Science* (11 January 2002), v. 295, p. 255. A short report on efforts to slow down light.

Seife, C., "The Intelligent Noncosmologist's Guide to Spacetime," *Science* (24 May 2002), v. 296, pp. 1418-1421. An excellent short history on the development of the theory of spacetime.

Seife, C., "Relativity Goes Where Einstein Sneered To Tread," *Science* (10 January 2003), p. 185. A short, accessible note on quantum entanglement and relativity.

Tegmark, M., "Measuring Spacetime: From the Big Bang to Black Holes," *Science* (24 May 2002), v. 296, pp. 1427-1433. A more advanced article on measurements of spacetime.

Wang, L., Kuzmich, A. and Dogariu, A., "Gain-Assisted Superluminal Light Propagation," *Nature* (20 July 2000), v. 406, pp. 277-279. Results of an experiment demonstrating superluminal light propagation in atomic cesium gas.

Will, C., "Gravitational Radiation and the Validity of General Relativity," *Physics Today* (October 1999), pp. 38-43. A discussion of how observations of gravity waves would test Einstein's theories.

Notes and Ideas

*Class time spent on material: Estimated:*______ *Actual:*______________

Related laboratory activities:

Demonstration materials:

Notes for next time:

Chapter 30: Quantum Physics

Outline

30-1 Blackbody Radiation and Planck's Hypothesis of Quantized Energy
30-2 Photons and the Photoelectric Effect
30-3 The Mass and Momentum of a Photon
30-4 Photon Scattering and the Compton Effect
30-5 The de Broglie Hypothesis and Wave-Particle Duality
30-6 The Heisenberg Uncertainty Principle
30-7 Quantum Tunneling

Summary

Chapter 30 discusses the beginning of quantum physics, which treats the behavior of microscopic systems. Quantization, the photon and its interactions with matter, wave-particle duality and its implications for subatomic particles, and the Heisenberg uncertainty principle are presented. The chapter ends with information on quantum tunneling.

Major Concepts

By the end of the chapter, students should understand each of the following and be able to demonstrate their understanding in problem applications as well as in conceptual situations.

- Blackbody radiation
  - Energy distribution
  - Wein's displacement law
- Quantization of energy
  - Planck's constant
- Photons
  - Energy
  - Momentum
- Photoelectric effect
  - Work function
  - Cutoff frequency
- Compton effect
- Wave-particle duality
  - The de Broglie hypothesis
- Heisenberg uncertainty principle
- Quantum tunneling

Teaching Suggestions and Demonstrations

This chapter contains many new ideas, some of which were very difficult for those who first developed them to accept. This material is conceptually complex. Stress the experimental proof of these ideas and do lots of example calculations.

Sections 30-1 through 30-3

Go over the definition of a **blackbody**. Students may have trouble with this concept. Have them think of a hot summer day. Would they rather walk barefoot on a white concrete sidewalk, or the black asphalt of the street? Why? Use their comments as a starting point for your description of a blackbody as a **perfect absorber** (see Figure 30-1) and a **perfect emitter**. Use Figure 30-2 to introduce **Wien's displacement law**. Talk about Conceptual Checkpoint 30-1 and the following section on temperature and color.

If you have time, step the students through the historical development of the **ultraviolet catastrophe**. (It's another of the great stories of physics!) It also helps the students understand how Planck could have proposed the **quantization of energy** when he didn't really think it was true. Go over the concepts of the **quantum number** and the energy $E = nhf$ of transitions between states. Work Example 30-1 in class.

Einstein was the first to apply the idea of quantization of energy to light. Work Exercise 30-2 and Example 30-2 in class before going on to the **photoelectric effect**. Discuss the differences between the classical model and Einstein's model of what happens in the photoelectric effect. Ask the students to tell you which model fits the data best. Go through Example 30-3 to show students how this analysis is done.

➲ **DEMO 30-1** Use a photoelectric cell to demonstrate the **photoelectric effect**. (See Resource Information.) Vary the intensity of the light and the frequency of the light to show students how each affects the current produced. (See Figures 30-4, 30-5, and 30-6.) Determine the **threshold frequency** for the cell and the **work function** of the metal.

Be sure to point out to the students that Einstein won the Nobel Prize for his work on the photoelectric effect, not relativity.

Discuss the properties of **photons**. Mention that the photon rest mass is zero, that photons travel at the velocity of light, and that photons carry momentum. The definition of **photon momentum** follows from the description of the relativistic energy. Go through this derivation and work Exercise 30-4. The applications of radiation pressure to space travel and microscopic manipulation are interesting.

Section 30-4

If you are short on time, you can skip this section entirely. If you have time to cover it, the **Compton effect** is a great example of the conservation of energy and momentum. Work out the derivation of Equation 30-15 in class and then go through Example 30-4.

Sections 30-5 through 30-7

Students are usually interested in the **de Broglie hypothesis**, particularly when they learn that de Broglie was a graduate student when he suggested it. Go over the de Broglie relation (Equation 30-16) and calculate the wavelength of several macroscopic objects (like baseballs) traveling at typical speeds. Discuss the reasons why we don't have to be concerned with the wave nature of large objects moving at normal speeds. Then work Active Example 30-2 for electrons.

Experimental proof of the de Broglie hypothesis first came from **electron diffraction** by crystals. Compare Figures 30-13 and 30-14. The similarities in these figures make a strong argument for the wave behavior of electrons. If you have time, work Example 30-5, which deals with **neutron diffraction**. Remind the students that all particles have a wave nature, just as all electromagnetic waves have a particle nature.

The **Heisenberg uncertainty principle** is difficult to explain at the introductory level. One way to make it seem more reasonable is to tie it to measurement. If we want to see a billiard ball rolling across a table at night, we can shine a flashlight beam on it. While we now know that the beam will create a radiation pressure on the ball, the force generated is negligible. The ball's path is undisturbed by the beam. We can calculate both the position of the ball and its momentum at some time t.

If we want to measure the position of an electron, what do we use as a probe? If we shine light on the electron, the photons in the light beam will interact with the electron and cause it to change position and/or momentum. It is possible to devise an experiment in which the position can be measured, or one in which the momentum can be measured. It is *not* possible to devise an experiment in which both can be measured simultaneously. In fact, it is not possible for both of these quantities to have precise values at the same time. Go through the qualitative argument given in Section 30-6 for the form of the Heisenberg uncertainty principle for position and momentum. Work Example 30-7. Discuss briefly the other forms of the uncertainty principle. Talk about Conceptual Checkpoint 30-4 and the paragraphs following on the results of changing the value of h.

Section 30-7 discusses **quantum tunneling**, a process that is responsible for alpha decay in nuclei and which has technological applications. The scanning tunneling microscope is a good example and is discussed thoroughly in the text. If you are pressed for time, you may choose to skip this section.

Resource Information

Transparencies

365. Figure 30-1	An ideal blackbody
366. Figure 30-2	Blackbody radiation
367. Figure 30-3	The ultraviolet catastrophe
368. Figure 30-4	The photon model of light
369. Figure 30-5	The photoelectric effect
370. Figure 30-6	The kinetic energy of photoelectrons
371. Figure 30-7	The Compton effect
372. Figure 30-8	Scattering from a crystal
373. Figure 30-9	Diffraction patterns
374. Figure 30-15	Diffraction pattern of electrons
375. Figure 30-16	Uncertainty in position and momentum
376. Figure 30-17	Optical tunneling
377. Figure 30-18	Operation of a scanning tunneling microscope

Physlet® Illustrations

30-1 Blackbody Spectrum
30-2 Photoelectric Effect
30-3 Scanning Tunneling Microscope

Suggested Readings

Birnbaum, J. and Williams, R., "Physics and the Information Revolution," *Physics Today* (January 2000), pp. 38-42. Discusses physical and economic limits to semiconductor devices and options for advancing computing involving quantum physics.

Chen, C. and Zhang, C., "New Demonstration of the Photoelectric Effect," *The Physics Teacher* (October 1999), p. 442. A modification of an introductory physics photoelectric-effect demonstration.

Fitzgerald, R., "What Really Gives a Quantum Computer Its Power," *Physics Today* (January 2000), pp. 20-22. An investigation of quantum computing with NMR.

Folger, T., "Quantum Shmantum," *Discover* (September 2001), pp. 36-43. An interesting article on the "'many worlds" interpretation of quantum mechanics.

Gottesman, D. and Lo, H., "From Quantum Cheating to Quantum Security," *Physics Today* (November 2000), pp. 22-27. Discusses the use of quantum phenomena to distribute code keys.

Horsewill, A., Jones, N. and Caciuffo, R., "Evidence for Coherent Proton Tunneling in a Hydrogen Bond Network," *Science* (5 January 2001), v. 291, pp. 100-103. An advanced article on evidence for quantum tunneling in a cyclic network of hydrogen bonds.

Isenberg, C., "Laser Diffraction Experiments with Pseudoliquids and Pseudosolids," *The Physics Teacher* (October 2000), pp. 411-413. Analog experiments using laser light to illustrate X-ray and neutron scattering.

Laloë, F., "Do We Really Understand Quantum Mechanics? Strange Correlations, Paradoxes, and Theorems," *American Journal of Physics* (June 2001), pp. 655-701. A comprehensive article on the present understanding of quantum mechanics.

Levi, B., "Researchers Stop, Store, and Retrieve Photons - or at Least the Information They Carry," *Physics Today* (March 2001), pp. 17-18. New developments in the study of photons.

Mermin, N., "The Contemplation of Quantum Computation," *Physics Today* (July 2000), pp. 11-12. A self-described old physicist and young philosopher presents thought-provoking issues of quantum computation.

Mohrhoff, U., "What Quantum Mechanics Is Trying to Tell Us," *American Journal of Physics* (August 2000), pp. 728-745. An interesting and novel interpretation of quantum mechanics; includes a thorough glossary of terms.

Smalley, E., "Future Tech: Hack-Proof Chatting," *Discover* (May 2002), pp. 26-27. A practical application of quantum principles.

Voss, D., "Quantum Engine Blasts Past High Gear," *Science* (18 January 2002), v. 295, p. 425. A short article on a theoretical "quantum afterburner" for heat engines.

See also the theme issue of the *American Journal of Physics*, March 2002, on quantum mechanics.

Materials

A photoelectric cell is available on its own or as part of the "Complete h/e System" (item AP-9370) from Pasco. The complete package includes the apparatus and instructions for performing photoelectric-effect experiments.

Notes and Ideas

Class time spent on material: Estimated:______ *Actual:____________*

Related laboratory activities:

Demonstration materials:

Notes for next time:

Chapter 31: Atomic Physics

Outline

Summary

Chapter 31 presents several early models of the atom, then introduces the quantum mechanical model of hydrogen. Data are presented that are inconsistent with the earlier models but which support the quantum model. The quantum model is extended to more complex atoms. The chapter ends with a discussion of various types of atomic radiation.

Major Concepts

By the end of the chapter, students should understand each of the following and be able to demonstrate their understanding in problem applications as well as in conceptual situations.

- Early models of the atom
  - Thomson model
  - Rutherford model
- Bohr model
  - Hydrogen spectrum
  - Assumptions
  - Orbits and energies
- The quantum mechanical model of hydrogen
  - Quantum numbers
  - Schrödinger equation
  - Electron probability clouds
- Multielectron atoms
- Atomic radiation

Teaching Suggestions and Demonstrations

By the time you reach this chapter, the end of the semester or quarter will be near. If you are running short on time, go over the models of the atom qualitatively. It is important to find time to discuss atomic radiation, especially if your course is aimed at promoting science literacy. Knowledge of the nature of radiation is key to making informed decisions about a variety of modern issues in medicine and society in general.

Sections 31-1 and 31-2

Discuss the two early models of the atom, the **Thomson model** and the **Rutherford model**, in terms of hypothesis, collection and interpretation of data, and reformulation of hypothesis. (See Figures 31-1 and 31-2.) The development of these models makes an exciting story; try to convey some of that sense of discovery to the students. Point out the flaws in each model and how the study of the flaws led to the next model.

One of the major problems with the Rutherford model of the atom was its prediction of a continuous atomic spectrum, since all charged particles traveling in a circular path are accelerating, and all accelerated charges emit radiation. **Line spectra** characteristic of individual elements were already known (see Figures 31-3 and 31-4) but not well understood. Go over the difference between **emission spectra** and **absorption spectra**. Be sure the students understand that the original **Balmer formula** was empirical, not derived. Go through Example 31-1 and calculate wavelengths of other lines in the **Balmer series**. Have the students refer to Figure 31-5 and discuss the **Lyman** and **Paschen series**.

➲ DEMO 31-1 Use a hydrogen spectral tube and the diffraction-grating glasses from Chapter 28 to let students view the Balmer series in hydrogen. (See Resource Information.) Have them identify the n value corresponding to each visible line. Let the students look at spectral lines from other elements as well. If you don't have a selection of spectral tubes, you can improvise by sending students across campus or out into the community to look at "neon" signs (which are mostly *not* neon) with the diffraction-grating glasses. Sodium and mercury parking-lot lights also work well!

Sections 31-3 and 31-4

The **Bohr model of the atom** is the model most people in modern society use. It is easy to imagine, explains many of the properties of atoms and uses a minimum number of assumptions. Go over the four basic assumptions and derive the equations for the radii of the **Bohr orbits** and the speed of an electron in these orbits. (See Figures 31-6 and 31-7.) Calculate the energies of electrons in the first several orbits and construct an **energy-level diagram**, as in Figure 31-8. Use the difference in energy of two levels to derive the Balmer equation. The derivation of the Balmer equation and the definition of the **Rydberg constant** in terms of fundamental constants were the triumphs of the Bohr model. Do Active Example 31-1. Remind students that the wavelengths calculated in this example can be **emitted** as an electron moves from a higher energy level to a lower one, or **absorbed** as an electron moves from a lower energy level to a higher one.

If you have time, go through Section 31-4, which connects the Bohr orbits to **standing de Broglie waves**. Use Figure 31-10 as an illustration and work Example 31-3. Discuss with the class how the success of the Bohr model led physicists to take matter waves seriously. Talk qualitatively about the **Schrödinger equation** and its importance in quantum mechanics.

Sections 31-5 and 31-6

Skim briefly through the **four quantum numbers** (n, , m , and m_s) required by the **quantum mechanical model** of the hydrogen atom. Pay particular attention to the **magnetic quantum number** and the **spin quantum number**, as these quantities will be very different from anything the students have seen before. The magnetic quantum number can be discussed as a quantized projection of the angular momentum vector. Be sure students understand that **spin** is an intrinsic property of the electron (and of other particles) and has no direct mechanical analog. Use the definitions of the quantum numbers to define a **quantum state** of hydrogen.

Talk qualitatively about the **wave function** of an electron in a hydrogen atom and about **electron probability clouds**. Use Figures 31-12 and 31-13. Point out that the electrons do not really travel in orbits, as in the Bohr model, but that there are certain regions of space where the electrons are more likely to be found. Don't expect the students to grasp all the details of quantum theory at this level. Go through Conceptual Checkpoint 31-2 and use it for class discussion.

Use the energy-level diagram for the hydrogen atom to begin talking about the **shell model** of atomic structure. Remind students that as you add more protons and neutrons to the nucleus and more electrons to the atom, the interactions between all of the parts become very complex. It is quite amazing that the simple shell model works as well as it does for so many atoms. Define the **Pauli exclusion principle** and discuss its implications for the filling of atomic **shells** and **subshells**. Go over the shorthand notation for the **electronic configuration** of an atom and do several examples. Chemistry students will be particularly interested in this topic.

➲ **DEMO 31-2** Get a large copy of the **periodic table** of the elements (you can borrow one from your colleagues in chemistry if you don't have one) and point out how the properties of the elements change with their electronic structure. Discuss why the table is set up in its present configuration.

Section 31-7

It is important to go through this section on **atomic radiation** at least qualitatively so that the students know the differences between the various types of radiation and the possible benefits and dangers of each. Be sure to talk about the different types of **X-ray** imaging available today. If you have time, the history of the medical use of X rays is very interesting. (See Resource Information.) **Lasers** have many more technological applications than X rays. Plan to spend some time discussing how a laser works and how the light emitted from a laser has special characteristics. Medical applications of lasers are many and varied. If you have a class full of biology majors or students in the pre-health professions, they will find this material fascinating.

➲ **DEMO 31-3** Bring a HeNe **laser** to class. Talk about the color of the light and how it is determined by the electronic structure of neon. Also discuss the fact that laser light is **coherent**. Ask the class to name uses for the laser that they see every day. Obtain a **hologram** that is visible with the laser, and give students a chance to view it.

Go through the mechanisms involved in **fluorescence** and **phosphorescence**. Discuss the relative energies of the photons given off as X rays with those given off in visible-light fluorescence.

Resource Information

Transparencies

378.	Figure 31-1	The plum-pudding model of an atom
	Figure 31-2	The solar system model of an atom
379.	Figure 31-4	The line spectrum of hydrogen
380.	Figure 31-5	The Lyman, Balmer, and Paschen series of spectral lines

381. Figure 31-6 A Bohr orbit
Figure 31-7 The first three Bohr orbits
382. Figure 31-9 The origin of spectral series in hydrogen
383. Figure 31-10 de Broglie wavelengths and Bohr orbits
384. Figure 31-12 Probability cloud for the ground state of hydrogen
Figure 31-13 Probability as a function of distance
385. Figure 31-17 Maximum number of electrons in an energy level
386. Figure 31-18 First eight elements of the periodic table
387. Figure 31-20 Designation of elements in the periodic table
388. Figure 31-22 X-ray spectrum
389. Figure 31-25 The helium-neon laser
390. Figure 31-27 Holography

Physlet® Illustrations

31-1 Rutherford Scattering
31-2 The Bohr Atom
31-3 Hydrogenic Atoms

Suggested Readings

Borbat, P., Costa-Filho, A., Earle, K., Moscicki, J. and Freed, J., "Electron Spin Resonance in Studies of Membranes and Proteins," *Science* (12 January 2001), v. 291, pp. 266-269. A review of electron spin resonance techniques used in biology.

Chu, S., "Cold Atoms and Quantum Control," *Nature* (14 March 2002), v. 416, pp. 206-210. An excellent overview article on the history of laser cooling of atoms, followed by five advanced articles.

Collins, D., "Video Spectroscopy - Emission, Absorption, and Flash," *The Physics Teacher* (December 2000), pp. 561-562. Uses a diffraction grating attached to a color video camera to display spectra.

Colson, W., Johnson, E., Kelley, J. and Schwettman, H., "Putting Free-Electron Lasers to Work," *Physics Today* (January 2002), pp. 35-41. Describes a new type of laser that is tunable over a wide range of wavelengths.

Day, C., "Ultrafast X-ray Diffraction Tracks Molecular Shape-Shifting," *Physics Today* (March 2001), pp. 19-20. A discussion of X-ray diffraction that can look at structural changes in molecules.

Lindley, D., *Boltzmann's Atom: The Great Debate That Launched a Revolution in Physics*, Free Press, NY, 2001. An interesting historical account of the work of Boltzmann.

Mould, R., *A Century of X-rays and Radioactivity in Medicine*, Institute of Physics Publishing, Bristol and Philadelphia, 1993. Wonderful photographs of early X rays and X-ray equipment.

Palmquist, B., "Interactive Spectra Demonstration," *The Physics Teacher* (March 2002), pp. 140-142. A creative method for modeling emission, absorption, and continuous spectra that involves throwing balls and jumping on chairs!

Peercy, P., "An Eye for Impurity," *Nature*, (25 April 2002), v. 416 pp. 799-80. A short article on the use of scanning transmission electron microscopy in locating the dopant atoms in semiconductors.

Peyser, L., Vinson, A., Bartko, A. and Dickson, R., "Photoactivated Fluorescence from Individual Silver Nanoclusters," *Science* (5 January 2001), v. 291, pp. 103-106. An advanced article on the storage and retrieval of data using nanoparticles and fluorescence.

Stevens, R., "A Simulation of Optical Polarization," *The Physics Teacher* (April 2000), pp. 222-223. A demonstration that can be used to model polarization states of light as well as particle spin.

See also the theme issue of the *American Journal of Physics*, March 2002, on quantum mechanics.

Materials

Eight different spectral tubes (items SE-9461 through SE-9468) and a spectral tube power supply and mount (item SE-9460) are available from Pasco.

The laser accessories kit from Fisher listed in the resource section of Chapter 28 includes a hologram. Kits for making holograms are also available from most major suppliers. See, for instance, the "Complete Holography Kit" (item S42800) from Fisher.

Ordering information for the diffraction-grating glasses is given at the end of Chapter 28.

Notes and Ideas

*Class time spent on material: Estimated:*______ *Actual:*____________

Related laboratory activities:

Demonstration materials:

Notes for next time:

Chapter 32: Nuclear Physics and Nuclear Radiation

Outline

Summary

Chapter 32 focuses on the nucleus, the center of the atom, and on elementary particles. The structure of the nucleus and its decay patterns are discussed. Biomedical applications of nuclear processes are presented. Basic categories of elementary particles are defined. The chapter ends with a brief description of unified field theory, the Grand Unified Theory, and cosmology.

Major Concepts

By the end of the chapter, students should understand each of the following and be able to demonstrate their understanding in problem applications as well as in conceptual situations.

- Nuclei
  - Constituents
  - Characteristics
  - Stability
- Radioactivity
  - Alpha decay
  - Beta decay
  - Gamma decay
  - Radioactive decay series
  - Activity
  - Half-life
  - Radioactive dating
- Nuclear binding energy
  - Fission
  - Fusion
- Applications of nuclear physics
- Elementary particles
  - Leptons
  - Hadrons
  - Quarks
- Unified forces and cosmology

Teaching Suggestions and Demonstrations

Just as in Chapter 31, there are issues in this chapter that students need to know about (at least qualitatively) so that they can be educated decision-makers for their own health and for the health of their communities. If you have only a single class to spend on these topics, try to describe nuclei, radioactivity, and nuclear processes qualitatively and spend time discussing the applications of nuclear physics to medicine and other fields.

Sections 32-1 through 32-3

These first three sections contain lots of new vocabulary. Students will generally know about protons and neutrons, and if they have already taken chemistry, they may be familiar with more of the terms. It is a good idea to make a poster, or write on the board in a semipermanent place, a list of the definitions of terms such as **nucleon**, **atomic number**, **neutron number**, and **mass number**. (See Table 32-1.) Go over **standard nuclear notation** ${}^{A}_{Z}X$, which will likely be unfamiliar to most students. Write several examples (like Exercise 32-1) on the board. Go over the definition of **isotopes** and remind the students that atoms with the same number of protons in their nuclei are the same element. Define the **atomic mass unit** u in terms of kilograms and MeV/c^2. Table 32-2 lists the masses of the proton, neutron, and electron in various units.

Work Example 32-1, which is Rutherford's estimate of the **radius of the nucleus**. Point out that the radius of the nucleus varies from element to element (Equation 32-4) and that the size of an average nucleus is on the order of one **fermi**. Go through Example 32-2 and calculate the average **density of a nucleus**. Try to give students an idea of how large this number is! Emphasize that although the radius of the nucleus depends on the mass number, the density does not.

If the protons in the nucleus all repel each other, what holds the nucleus together? Ask the students this question and use their answers to introduce the **strong nuclear force**. Go over the properties of this force. Discuss the competing influences of strong nuclear attraction and electrostatic repulsion and their effects on the stability of the nucleus. You can then begin to talk about the **radioactive decay** of unstable nuclei. Discuss the main types of particles (**alpha particles**, **electrons**, **positrons**, and **photons**) that are emitted during the decay of a nucleus. Go over the properties of the positron and point out that electrons and positrons are often grouped together and called **beta particles**. Mention that other particles can be emitted by a nucleus, but that these are the most common. Discuss the **mass defect** and how it contributes to the large amount of energy generated in nuclear decays. Conceptual Checkpoint 32-1 is a nice exercise to help students connect this material to what they already know.

➲ **DEMO 32-1** If you have access to a **Geiger counter** or other radiation detector and some low-level sources of radiation (alpha, beta, and gamma), bring them to class. (See Resource Information.) Use a source of each type of radiation in turn. Start with the detector close to the source and slowly move it farther and farther away. This will demonstrate the **range** of the radiation in air. Then hold the detector at a constant distance and place barriers between the source and the detector. Use paper, aluminum sheets, and lead. Begin with one piece, then add more. This will demonstrate the **penetrating ability** of the various types of radiation. Make a chart on the board and begin to talk about the three main types of decay.

Point out that an **alpha particle** is just a helium nucleus, which has a particularly stable structure. Discuss what quantities change in a nucleus when it emits an alpha particle. Be sure that the students understand that the **parent nucleus** and the **daughter nucleus** are different elements. Go through Example 32-3 and

Conceptual Checkpoint 32-2 in class and talk about how the excess energy is carried off in alpha decay. The Real World Physics application on smoke detectors following Conceptual Checkpoint 32-2 will be of interest to the students.

Go over the basic reaction (neutron conversion to a proton) involved in **beta decay** and work Example 32-4. Talk about the distribution of kinetic energy in beta decay and how measurement of this energy led to the discovery of the **neutrino** and the **antineutrino**. (See Figure 32-3.) Point out that in **gamma decay** the parent and daughter nuclei are the same isotope.

Choose one **radioactive decay series** (Figure 32-4 describes the decay of ${}^{235}_{92}U$) and go through it with the students. Comment on how decay series' can be used to determine an estimate of the age of the Earth. Define the **activity** of a sample and its units, the **curie** and the **becquerel**.

As you begin to talk about **half-life**, point out that radioactive decay is a statistical process. We can make predictions about when a percentage of the nuclei in a sample will decay, but we cannot predict when a particular nucleus will decay. You can discuss this in relation to predictions on the average age at which people get married, or in relation to life insurance studies of death rates in the general population. If most of the students in your class are comfortable with logarithms, go through the derivation of Equation 32-9.

➲ **DEMO 32-2** Give each student in the class the same number of pennies. (This works best if the total number of pennies is close to 1000. You can use any kind of disk with two different sides. Pennies are common and easy to obtain.) The total number of pennies for the class will be N_0. Have all of the students flip the pennies at the same time. Each time they flip, one half-life has passed. Have them report the number of heads they get and discard the pennies that came up tails. Repeat this process about 10 times, or until most of the students have no pennies left. Record the results of each flip on the board or overhead. Plot the result on rectangular graph paper and on semi-log paper. (Make a transparency of a piece of each type of graph paper and do this on an overhead.) The exponential decay curve is quite dramatic and usually quite smooth! It will be easy to see that the number of pennies left decreases by half in equal amounts of time.

Talk about the relationship between **activity** and half-life. Do Active Example 32-3 in class. The students will find **carbon-14** dating interesting. Work Example 32-5 and talk about the limits of the procedure.

Sections 32-4 through 32-7

Begin this section by discussing the relationship between the mass defect and the **binding energy** of a nucleus. Then define the binding energy per nucleon and go over Figure 32-9. Point out the **region of nuclear stability**, the fission region, and the fusion region. Talk about the **fission** and **fusion** processes. Be sure to discuss the energy per nucleus released in each process and the reaction products. If you have time, go over **chain reactions** and the basic structure of a **nuclear reactor**. Do try to find time to talk about the **proton-proton cycle** and the Sun.

All students find a discussion of the **biological effects of radiation** interesting. The units tend to be confusing, so go over the definitions of the **roentgen**, the **rad**, the **RBE**, and the **rem** carefully. Work Exercise 32-4 in class. There are Web sources of average radiation doses for common activities. (See Resource Information.) Several medical applications of nuclear physics are given at the end of Section 32-7. If you have time, talk to the class about risks versus benefits for these procedures.

Sections 32-8 and 32-9

Point out to the students that of all the **elementary particles** they have studied so far, only electrons and neutrinos are fundamental particles. Go over the definition of a **lepton** and talk about which particles fall into this category. (See Table 32-6.) Then briefly discuss **hadrons** and **quarks**. (See Table 32-7.) If you have time, give a general overview of the **four fundamental forces** and the idea of a **Grand Unified Theory**. Students will run into these ideas in the popular press; it is nice if they have some sense of what they mean.

Resource Information

Transparencies

391. Table 32-2	Mass and Charge of Particles in the Atom
392. Figure 32-1	N and Z for stable and unstable nuclei
393. Figure 32-2	Alpha decay of uranium-238
394. Figure 32-4	Radioactice decay series of U-235
395. Figure 32-7	Activity of carbon-14
396. Figure 32-8	The concept of binding energy
397. Figure 32-9	The curve of binding energy
398. Figure 32-10	Nuclear fission
399. Figure 32-12	The quark composition of mesons and baryons
400. Figure 32-13	The evolution of the four fundamental forces

Physlet® Illustrations

32-1 Radiation in a Magnetic Field
32-2 Half-Life
32-3 Nuclear Chain Reaction

Suggested Readings

Aberg, S., "Weighing Up Nuclear Masses," *Nature* (30 May 2002), v. 417, pp. 499-501. A short article on the discrepancy between calculated and measured nuclear masses.

Brumfiel, G., "Cosmology Gets Real," *Nature* (13 March 2003), pp. 108–110. An accessible update on modern cosmology.

Drell, S., "Andrei Sakharov and the Nuclear Danger," *Physics Today* (May 2000), pp. 37-41. A look at some of the ideas of 1975 Nobel Peace prizewinner Andrei Sakharov as they apply in the post–cold war world.

Feder, T., "Food Meets Physics at Irradiation Facility," *Physics Today* (April 2002), p. 24. A short article on food irradiation studies.

Franklin, A., "The Road to the Neutrino," *Physics Today* (February 2000), pp. 22-28. Historical description of the discovery and subsequent study of the neutrino.

Garrison, E., "Physics and Archeology," *Physics Today* (October 2001), pp. 32-36. Outlines important contributions of physics to archeology, including radioisotope dating as well as many newer physics-based techniques.

Haxton, W. and Holstein, B., "Neutrino Physics," *American Journal of Physics* (January 2000), pp. 15-32. Technical and thorough treatment of the basic concepts of neutrino physics.

Lubkin, G., "Nobel Prize to 't Hooft and Veltman for Putting Electroweak Theory on Firmer Foundation," *Physics Today* (December 1999), pp. 17-19. Discussion of the recipients of the 1999 Nobel Prize in Physics and their work.

Macklis, R., "Scientist, Technologist, Proto-Feminist, Superstar," *Science* (1 March 2002), v. 295, pp. 1647-1648. An excellent short article on Marie Curie.

Mak, S., "Radioactivity Experiments for Project Investigation," *The Physics Teacher* (December 1999), pp. 536-539. Discusses the use of potassium chloride in half-life estimation and other radioactivity experiments.

Riordan, M., Rowson, P. and Wu, S., "The Search for the Higgs Boson," *Science* (12 January 2001), v. 291, pp. 259-260. A short article on the search for the Higgs boson, an important piece of the Standard Model.

Sawicki, M., "A Note on Quantum Teleportation," *The Physics Teacher* (November 1999), p. 499. An analogy illustrating the essence of quantum teleportation experiments.

Seife, C., "Atom Smasher Probes Realm of Nuclear 'Gas'," *Science* (25 January 2002), v. 295, pp. 603-604. Discusses new developments in the liquid drop model of the nucleus.

Seife, C., "Muon Experiment Challenges Reigning Model of Particles," *Science* (9 February 2001), v. 291, p. 958. A short article on experimental evidence of the potential failure of the Standard Model.

Steiger, W., "A Radioactive Tracer in Medicine," *The Physics Teacher* (October 1999), pp. 408-409. Of particular interest to students in the pre-health professions.

Stone, R., "Living in the Shadow of Chornobyl," *Science* (20 April 2001), v. 292, pp. 420-426. A description of the Chornobyl accident and its long-term aftermath, with interviews.

Tigner, M., "Does Accelerator-Based Particle Physics Have a Future?," *Physics Today* (January 2001), pp. 36-40. A discussion of fundamental questions in particle physics and the experimental facilities that can address them.

The Environmental Protection Agency Division of Radiation Protection is an excellent source of information and data on radiation exposure. The Web site address is: http://www.epa.gov/radiation/.

Materials

A basic radiation meter is available from Fisher Science Education, model number S44126B.

A radioactive source set is also available from Fisher Science Education, model number S440975.

A basic nuclear lab station, including radioactive sources, model number S46860IBM, and an intermediate nuclear lab station, model number S46863IBM, are also available from Fisher Science Education. These stations come in MAC versions as well.

Notes and Ideas

Class time spent on material: Estimated:______ Actual:_____________

Related laboratory activities:

Demonstration materials:

Notes for next time:

Just in Time Teaching Notes

Gregor Novak
Andrew Gavrin
Indiana University–Purdue University, Indianapolis

---

to accompany

Physics

Second Edition

James S. Walker
Western Washington University

Just in Time Teaching Notes

to accompany *Physics, second edition* by James S. Walker

Table of Contents

Introduction to Just-in-Time Teaching using *Physics* by James S. Walker

Many of the resources associated with this text are designed for use with a method known as "Just-in-Time Teaching," or "JiTT" for short. In particular, the Warm-Up questions and the Puzzles were written for this purpose. These exercises appear on the Walker *Physics, 2nd edition* Course and Homework Management Systems (WebCT, BlackBoard, CourseCompass, PHGradeAsssist) and in the Student Study Guide. In order to help you implement the JiTT strategy, this manual includes teaching notes on many of the Warm-Ups and Puzzles, including sample student responses, and references to the corresponding sections of the Walker text. What follows is a brief outline of the JiTT method, its underlying philosophy, and its impact on student learning. For a more detailed presentation of the JiTT method, please visit the JiTT Web Site, http://webphysics.iupui.edu/jitt/jitt.html, or obtain a copy of the JiTT book, Just-in-Time Teaching: Blending Active Learning with Web Technology, by Gregor M. Novak, et al., published by Prentice Hall, Inc., 1999.

What Is Just-in-Time Teaching?

The Just-in-Time Teaching (JiTT) strategy is aimed at many of the challenges confronting instructors and students in today's classrooms. Student populations are diversifying; in addition to the traditional nineteen-year-old recent high school graduates, we now have a kaleidoscope of "non-traditional" students: older students, working part-time students, commuting students, and, at the service academies, military cadets. At a minimum, these students face time-management challenges. They also need motivation and encouragement to persevere in what for many is a bewildering, unfamiliar task. Consistent friendly support often makes the difference between graduating and dropping out.

Education research has made us more aware of learning style differences and of the importance of passing some control of the learning process over to the students. Active learner environments yield better results, but they are harder to manage than lecture-oriented approaches.

To confront these challenges, the Just-in-Time Teaching strategy pursues three major goals:

1. To maximize the efficacy of the classroom session, where human instructors are present.
2. To structure the out-of-class time for maximum learning benefit.
3. To create and sustain team spirit. Students and instructors work as a team toward the same objective: to help all students pass the course with the maximum amount of retainable knowledge.

Although Just-in-Time Teaching makes heavy use of the Web, it is not to be confused with either distance learning (DL) or with computer-aided instruction (CAI.) Virtually all JiTT instruction occurs in a classroom with human instructors. The Web materials, added as a pedagogical resource, act primarily as a communication tool and secondarily as content provider and organizer.

JiTT Web pages fall into three major categories:

1. Student assignments in preparation for the classroom activity. Warm-Ups and Puzzles fall into this category.
2. Enrichment pages. We title these pages "What is Physics Good For?" ("GoodFors") for short. These are short essays on practical, everyday applications of the physics at hand, peppered with URL links to interesting material on the Web. These essays have proved to be an important motivating factor in introductory physics service courses, where students often doubt the current relevance of classical physics, developed hundreds of years ago.
3. Stand-alone instructional material, such as simulation programs and spreadsheet exercises.

What Are JiTT Warm-Ups and Puzzles?

Warm-Ups and Puzzles are short Web-based assignments, which prompt the student to think about a physics-related topic and answer a few simple questions prior to class. It can be seen from examples in the following pages that some of these questions, when discussed fully, often have complex answers. We expect the students to develop the answer as far as they can on their own, and we finish the job in the classroom. These assignments are due just a few hours before class time. The responses are collected electronically and scanned by the instructor in preparation for class. They become the framework for the classroom activities that follow. In a typical application, sample responses are duplicated on transparencies and taken to class. In an interactive session, built around these responses, the lesson content is developed.

Students complete the Warm-Up assignments before they receive any formal instruction on a particular topic. They earn credit for answering a question, substantiated by prior knowledge and whatever they managed to glean from the textbook. The answers do not have to be complete or even correct.

Puzzle exercises are assigned to students after they have received formal instruction on a particular topic. They serve as the framework for a wrap-up session on a particular topic.
The Warm-Ups, and to some extent the puzzles, are designed to deal with a variety of specific issues. These can be roughly categorized as follows.

- Developing Concepts and Vocabulary
- Modeling – Connecting Concepts and Equations
- Visualization in General and Graphing in Particular
- Estimation, Getting a Feel for Magnitudes
- Relating Physics Statements to "Common Sense"
- Understanding Equations – the Scope of Applicability

In preparing Warm-Up assignments for an upcoming class meeting, we first create a conceptual outline of the lesson content. This task is similar to the preparation of a traditional passive lecture. As we work on the outline, we pay attention to the pedagogical issues that we need to focus on when in front of the class. Are we introducing new concepts and/or new notation? Are we building on a previous lesson, and if so, what bears repeating? What are the important points we wish the students to remember from the session? What are the common difficulties typical students will face when exposed to this material? (Previous classroom experience and education research can be immensely helpful here.) Once this outline has been created, we create broadly based questions that will force students to grapple with as many of the issues as possible. We are hoping to receive, in the student responses, the framework on which we build the in-class experience. When students leave a JiTT classroom, they will have been exposed to the same content as in passive lecture with two important added benefits. First, having completed the Web assignment very recently, they will be ready to actively engage in the classroom activities. Second, they will have a feeling of ownership since the interactive lecture is based on their own wording and understanding of the relevant issues. To close the feedback loop, the "give" and "take" in the classroom suggests future Warm-Up questions that will reflect the mood and the level of expertise in the class at hand. Thus, from the instructor's point of view, the lesson content remains pretty much the same from semester to semester. From the students' perspective, however, the lessons are fresh and interesting, with a lot of input from the class.

Gregor Novak
Andrew Gavrin
Indiana University–Purdue University, Indianapolis

Chapter 1: Introduction

WARM-UPS

1. Do you believe that the metric system is superior to the previous systems of measure, such as the everyday system used in the United States? Whatever your answer, what arguments would you use to persuade a person who has a different opinion?
Section 1-2

This question asks the student to think beyond the technical definitions and consider the social and economical consequences of scientific and technical issues. This makes an otherwise dry subject (at least to the average student) a bit more interesting.

Some advantages of the SI system: It conforms with international practice; the units and subunits are related via powers of 10, the standards are defined in terms of reproducible physical phenomena. A short-term disadvantage: A change would require some industries to re-tool.

2. Estimate the number of seconds in a human "lifetime." We'll let you choose the definition of "lifetime." Do all reasonable choices of "lifetime" give answers that have the same order of magnitude?
Section 1-6

Here is a chance to discuss estimation and order-of-magnitude via an easy-to-understand example.

The order of magnitude estimate: 10^9 *seconds*
$70\ yr = 2.2 \times 10^9\ s$ $100\ yr = 3.1 \times 10^9\ s$ $50\ yr = 1.6 \times 10^9\ s$

3. Which is a faster speed, 30 mph or 13 meters/second? Describe in words how you obtained your answer.
Section 1-5

30 mph = 13.3 m/s, slightly faster than 13 m/s.
The question forces the student to describe the unit-conversion process in words rather than formulas.

4. Estimate how many 20 cm x 20 cm tiles it would take to tile the floor and three sides of a shower stall. The stall has a square floor, which is 16 square feet, and five-foot walls.
Section 1-6

Total area to tile is 16 sq ft + 3 x 20 sq ft = 76 sq feet ~ 11,000 sq in. A tile covers 400 sq cm ~ 64 sq in. The area can be covered by ~ 172 tiles. Further discussion: Can one do this without cutting any tiles? If tiles have to be cut, do we need more than 172 tiles?

PUZZLES

"Where are you?"

The standard geographical coordinates of Chicago are:

Latitude: 41 degrees 50 minutes.
Longitude: 87 degrees 45 minutes.
What are the *x,y,z* coordinates of Chicago in a coordinate system centered at the center of the Earth with the *z*-axis pointing from the South Pole to the North pole, and the *x*-axis passing through the zero longitude meridian pointing away from Europe into space?
Please answer this in words, not equations, briefly explaining how you obtained your answer.
Section 2-1

The student is asked to give a verbal summary of data conversion from a spherical coordinate system to the Cartesian coordinate system.

Spherical polar coordinates: radius $= r$, *latitude* $= \theta$, *longitude* $= \phi$
Converting to Cartesian coordinates:
$x = r\cos\theta\cos\phi$
$y = r\cos\theta\sin\phi$
$z = r\sin\theta$

In words:
Draw the radius of the sphere from the center to the point on the surface representing Chicago.
The z-coordinate of Chicago is the component of the radius vector along the polar axis, the z axis. The radius vector makes a 48° 10'' angle with z axis. The z component of the radius is thus R_E cos(*48° 10''*) $=0.67R_E$.
The x and y coordinates of the point on the sphere are the x and y components of the projection of the radius vector on the equatorial plane.
The length of the projection vector is R_E *cos(41°50'').*
The x component is R_E *(cos 41°50'')(sin 87°45''); the y component is* R_E *(cos 41°50'')(cos 87°45'').*

Here are some sample (unedited) student responses that provide useful starting points for discussion:

a. *Using the coordinate system given in the puzzle, a displacement vector C begins at the origin and ends at Chicago. The magnitude of this vector is equal to the radius of the Earth (3963 miles). The latitude of Chicago converts to an angle with respect to the xy plane, and the longitude of Chicago converts to an angle with respect to the x axis. Then, trigonometric formulas can be used to determine the x,y, and z components of the vector C. These components are the x,y, and z coordinates of Chicago (given in miles)*
(x,y,z) = (115.9, -2950.7, 2643.0)

b. *All calculations were based on the given longitude and latitude and radius of the Earth (1982.17). The given latitude provides the Z component. Since Chicago is on the Earth's surface, I knew the distance to the origin.(1982.17) I used trigonometry to find Z. Z also determines the distance from Chicago to the origin on the ZY plane. Since I now had the distance of the projected Z value, I could work in two dimensions. I used the given longitude to turn an angle from X to Y. Using this angle and the projected distance on the XY plane, I solved for X and Y. To check the answer, I added the X,*

Y,and Z vectors and obtained the resultant vector from the origin to Chicago. It was equal to the radius of the Earth, so the answer made sense.
(X,Y,Z)=> (58,-1476,1309) (in miles)

c. *I let the z-axis point from the South Pole to the North Pole and the x-axis passing through the zero longitude meridian pointing away from Europe. The prime meridian then lies in the x,z plane, and the equator lies in the x,y plane. The center of the earth lies at the origin. I let the radius of the earth be equal to the distance from the center of the earth to Chicago which is 6.38x10^6m. I converted the longitude and latitude to degrees by letting 60 minutes equal 1 degree. By using trigonometric strategies such as 'sin's' and 'cos's' with these newly converted degrees, I found the (x,y,z) to be as follows: (1.86x10^5m, 4.75x10^6m, 4.26x10^6m).*

d. *I first went about the problem by converting the given into spherical coordinates. In examining a globe, I found that if X is pointing away from Europe, the Longitude needed to be recalibrated to a new Zero axis running through the 180 degree Longitude mark, on the other side of the world from the Prime Meridian. I looked up the radius of the Earth, approximately 6440 Km. This gave me the spherical coordinates (in radians) of (6440Km, 1.61, 0.73). Converting this into rectangular coordinates, I ended with the answer in (X, Y, Z) of (-168.3 Km, 4291.3 Km, 4798.9 Km).*

e. *To answer this question, many things must be assumed, since the earth is not completely spherical, we must assume that it is. By knowing that the equation of a sphere and drawing a 3 d picture we can see that the latitude creates a angle along the z axis. The longitude creates an angle on the x axis. By picking an arbitrary point (A) on the surface of the sphere and drawing a line from the center of the sphere to A we create a line that is in the zy plane. From point A a line can be drawn to the x axis to form a line perpendicular with the x axis. This can then be used to draw a right triangle where the x coordinate can be found. By dropping a line from point A down to the xy plane and then connecting it to the point drawn on the x axis we create another right triangle where the y value can be determined. The z value can be found by determining the height of the triangle formed in the zx plane. In order to do this, the radius of the earth had to be supplied. By determining the radius was 6.38E6m, we found that the x,y,and z coordinates were 1.67E5i + 4.25E6j + 4.75E6k*

Chapter 2: One-Dimensional Kinematics

WARM-UPS

1. During aerobic exercising, people often suffer injuries to knees and other joints due to HIGH ACCELERATIONS. When do these high accelerations occur?
Section 2-4

The short answer is that high accelerations occur when the person's feet hit the floor. This question prepares the student for a discussion of definitions of kinematics quantities. Many students find it difficult to grasp the idea of rate of change. This example involves abrupt stops (high rate of change of magnitude), abrupt change in direction (calling attention to the vector nature of kinematics quantities), and, implicitly at least, it prepares the ground for associating accelerations with forces.

When we introduce the concept of acceleration for the first time, the student already "knows" what we mean by the word accelerate. However, in the restricted sense of physics terminology, acceleration is defined as the rate of change of either the speed or the direction of motion. Thus, accelerations occur when we speed up (the common notion), but also when we slow down or change direction. Dealing with rates is notoriously difficult. It has to be explicitly, and frequently, pointed out to beginning students that very high accelerations can occur at very low speeds if the rate of change is high, i.e., if the time interval during which the change occurred is very short.

Here are some sample (unedited) student responses that provide useful starting points for discussion:

a. *These high accelerations occur when you start and stop the exercising because the initial and final speed is zero. While exercising, the speed of bending joints is pretty much constant; therefore, the acceleration is zero.*

b. *Injuries to the knee occur when the foot makes contact with the floor. The acceleration swiftly goes to zero and then back up again in the opposite direction that it was just going. This causes great stress on the knee and is when most of the injuries occur.*

c. *The greatest acceleration would be when the motion is being least resisted by the force of gravity; for example, if someone is doing step aerobics, the greatest acceleration would be when the person's leg is moving down toward the step or down toward the floor.*

2. Estimate the acceleration you subject yourself to if you walk into a brick wall at normal walking speed. (Make a reasonable estimate of your speed and of the time it takes you to come to a stop.)
Section 2-4

This can be answered by estimating the time it takes to stop you or by estimating the distance the body moves as you compress against the wall. The second estimate is easier and more intuitive, but it does not use $a = dv/dt$, which in our experience is what most students use at this point. We often use this kind of question to discuss the difference between a very short time and an instant (as in instantaneous velocity). Perceptive students may remark on this, as in the comment below. This kind of comment gives us an opening to bring up the issue.

Here are some sample (unedited) student responses that provide useful starting points for discussion:

a. *Assuming an initial velocity of 1 m/s, I estimate the acceleration experienced when walking into a brick wall to be approximately -50 m/s^2. This would occur in about .02 seconds.*

b. *'Make a reasonable estimate of your speed and of the time it takes you to come to a stop.' The time it takes me to come to a stop. This seems a bit confusing to me...wouldn't it be more or less instantly? Once you hit that brick wall you aren't going any further, no? I found the time by figuring when you'd 'start' walking into the walk till the point you wouldn't be doing any more walking. A ball thrown at the brick wall immediately stops or is the amount of time between contact, and the start of the fall considered 'stopping time?'*

3. A man drops a baseball from the edge of a roof of a building. At exactly the same time, another man pitches a baseball vertically up toward the man on the roof in such a way that the ball just barely reaches the roof. Does the ball from the roof reach the ground before the ball from the ground reaches the roof, or is it the other way around?
<u>Section 2-5</u>

They reach their respective destinations at the same time. Many students have trouble with this question. Some get this correctly but their logic is often faulty or incomplete. We want them to answer with a short essay, but we want the verbal supported by an appeal to mathematical arguments.

Here are some sample (unedited) student responses that provide useful starting points for discussion:

a. *The ball from the roof would reach the ground first because the ball going upwards has gravity acting on it.*

b. *The ball dropped would reach the ground at exactly the same time as the one shot because they are traveling the same distance with the same acceleration acting upon them. Unless, of course, the balls collided, in which case the ball shot would never reach the edge, ensuring victory for the dropped ball.*

c. *I say the ball from the roof reaches the ground faster because the ball shot from the man below is slowing down due to the effects of gravity. The falling ball is picking up speed as it reaches the ground.*

d. *The ball from the roof reaches the ground first. The ball falls at a constant rate of 9.8m/s2 and since the ball from the ground barely reaches the roof, it must slow down on its way up. The ball from the roof never slows down on its way to the ground.*

e. *The ball shot from the ground will reach the roof before the ball dropped from the roof reaches the ground. This is because the acceleration of a freely falling object is constant while the acceleration of the ball shot up from the ground is increasing.*

f. *Neither, the amount of time for the two balls to reach their respective destinations would be the same. For example, if the building was exactly 9.8m tall, it would take the ball being dropped exactly one second to reach the ground, and due to the downforce of gravity being -9.8m/s, if the ball is shot at 9.8m/s then the ball will go exactly 9.8m and it would also take one second.*

4. Estimate the time it takes for a free-fall drop from 10 m height. Also estimate the time a 10-m platform diver would be in the air if he takes off straight up with a vertical speed of 2 m/s (and clears the platform, of course!)
<u>Section 2-5</u>

Estimated time for the drop is straightforward, about 1.4 s. The quickest way to estimate the time in the second case is via average velocity, which is about - 6 m/s. Having made the estimate, students then check this value by actually solving the quadratic equation.

PUZZLES

"Round-Trip"

An airplane flying at constant air speed, due east from My City to Your City, in calm weather (no wind of any kind) would log the same flying time for both legs of the trip. Suppose the same trip is taken when there is wind from the west. How would the total (round-trip) time in windy weather compare with the total time in calm weather? Please answer this in words, not equations. Explain briefly how you obtained your answer.
<u>Section 2-2</u>

Let the air speed of the airplane be v. In calm air, the air speed is also the ground speed. In calm air the round-trip flight time is then 2D/v, where D is the distance between the cities. Denote the wind speed with v_w. *The ground speeds are then* $v + v_w$ *and* $v - v_w$, *respectively. Thus, the time with the wind* $t_1 = D/(v + v_w)$ *and the time against the wind* $t_2 = D/(v-v_w)$. *The round trip time now becomes* $D/(v + v_w) + D/(v-v_w) = 2D/(v^2 - v_w^2)$. *Check extreme cases: For* v_w *this reduces to 2D/v, as it should. For* $v_w = v$ *the time is infinite.*

Here are some sample (unedited) student responses that provide useful starting points for discussion:

a. *Logically speaking, if there is a constant wind from the west going to and from, that means that there is a head wind going one way and a tail wind coming back. If this is the case, and the force that the plane exerts is the same for both flights (with wind and without), then the time comparison would be equal.*

b. *The round trip time would be the same if wind speed was constant in both magnitude and direction, the time lost against the wind going west would be made up coming east.*

c. *The total time for the second trip would take longer on the first part of the trip because the plane is not only slowed down by the velocity of the wind but to stay on course the pilot must change his heading or he will be blown off course. On the return trip if the velocity of the wind is the same the plane will be pushed faster than both the first trip and the trip going to St. Louis with the wind. Then one would have to assume that if both the trip there and the return trip were added the time would be the same as the first trip all together.*

d. *Assuming the same wind velocity for each leg, the flight times should still be the same. The plane would be slowed by x minutes en route to St. Louis, but then would be sped up by x minutes on the return trip.*

e. *Travelling with a headwind would lengthen the travel time. Coming back with a tailwind, would shorten the travel time. But if the velocity cannot be increased the total travel time would still be longer than for the totally round trip in calm weather. The plane in windy weather cannot make up the time lost due to headwind.*

f. *The total round trip would take longer with the west wind. This is because the headwind would be acting upon the plane longer than the tailwind. For example, if the headwind component on the outbound leg was 50 knots, the airspeed of the plane was 250 knots, and the distance from Indy to St. Louis was 200 nautical miles, then the outbound trip would take one hour. The return trip would take 40 min., for a round trip time of 1 hr. 40 min. This is longer than the 1 hr. 36 min. it would take to fly the 400 nautical mile round trip in no wind conditions. The 50 knot tailwind acted upon the plane for 1 hour, but the 50 knot tailwind only acted upon the plane for 40 min.*

Chapter 3: Vectors in Physics

WARM-UPS

1. Which of the following are vector quantities?
a) tension in a cable
b) weight of a rock
c) volume of a barrel
d) temperature of water in a pool
e) drift of an ocean current
Section 3-2

Students may find the answer b) obvious, but they may have more trouble with a) and e). After having thought about these examples, they are ready to talk about coordinate systems, components of vectors, and vector algebra.

2. Is there a place on the Earth where you can walk due south, then due east, and finally due north, and end up at the spot where you started?
Section 3-3

The answer is yes, the North Pole. This is not obvious to most students. It is an opportunity to talk about non-cartesian coordinate systems.

3. Is it possible for the magnitude of the vector difference of two vectors to exceed the magnitude of the vector sum?
Section 3-3

Again, after students have experimented with this question, we are ready in class to determine that
sum > diff if the angle between the vectors < 90˚.
sum = diff if the angle between the vectors = 90 .
sum < if the angle between the vectors > 90.

4. Reversing the algebraic sign on the velocity vector reverses the direction of motion.
Is the same statement true for the acceleration vector?
Section 3-5

No. The direction of the acceleration vector is determined by the direction of the vector difference of two velocity vectors. For students this is a difficult notion. In class we start with straight-line motion. This is what most student responses will deal with. The sign of the acceleration vector relative to the sign of the velocity vector tells us whether the object is speeding up or slowing down. Eventually we bring in the acceleration vector when the direction of motion is changing, preparing the ground for the introduction of circular motion.

PUZZLES

"Sharing the Burden"

The helpful twin brothers are swinging their little 30-lb sister on a rope.
Is it possible to arrange the rope so that each of the twins holds up 30 lbs?
<u>Section 3-3</u>

Denote the angle between the ropes by θ. The vector sum of the forces on the baby must be zero. If each sibling holds up 30 lbs, the sum of the up components of the vertical forces must be 2(30 lbs)sin(θ/2). The down force is 30 lbs.
Therefore 2(30 lbs)sin(θ/2) - 30 lbs = 0
sin(θ/2) = 1/2
(θ/2) = 30°
θ = 60 °.

Chapter 4: Two-Dimensional Kinematics

WARM-UPS

1. To attain the maximum range, a projectile has to be launched at 45 degrees if the landing spot and the launch spot are at the same height (neglecting air-resistance effects). Explain in a few sentences how the relation between the vertical and the horizontal components of the initial velocity affects the projectile range.
Section 4-2

If the launch point and the landing point are at the same elevation, the vertical component of the launch velocity will determine the time of flight; the horizontal component and the time of flight will determine the range. For maximum range this relation has to be optimized. If the launch and landing points are at different elevations, the optimization process becomes more difficult. The optimum angle now depends on the elevation difference.

2. On the Moon the acceleration due to gravity is about one-sixth that on Earth. If a golfer on the Moon imparted the same initial velocity to the ball as she does on the Earth, how much farther would the ball go?
Section 4-2

If the initial velocity stays the same, the vertical and the horizontal components of the ball's velocity will be the same. The range will change because the time in the air will be longer with g being less. Since the time in the air depends on the square root of g, the range will be about 2.5 times longer.

3. Three swimmers start across the river at the same time. They all swim with the same speed relative to the water. Swimmer A swims in a direction perpendicular to the current. Swimmer B swims slightly upstream along a line that makes an 80-degree angle with the shore. Swimmer D swims slightly downstream along a line that makes an 80-degree angle with the shore. Which of the three swimmers reaches the opposite shore first?
Section 4-1

As seen from the typical student answers below, the notion of components does not come easy when some common sense notions interfere. Some common mistakes: shortest distance, shortest time – with stream, faster net speed, shortest time.

Here are some sample (unedited) student responses that provide useful starting points for discussion:

a. *The swimmer who starts on a path slightly upstream will reach the other side first. The other two swimmers will be pushed downstream by the current thus making them travel a longer distance.*

b. *Swimmer A will reach the opposite side first. The velocity of the stream only causes the swimmers to be pushed downstream. In does not affect their velocity toward the opposite shore. Since swimmer B and C are not taking 'direct' routes, they will arrive after swimmer A.*

c. Since all the swimmers are at the same speed, Swimmer B will get to the other side first. He will have the shorter distance to swim. With the other swimmers swimming at an angle, they have a farther distance. The shortest distance between two points is a straight line.

d. They will reach the shore at the same time, since they are all going the same speed. However their positions and the distance they traveled will differ.

4. The pilot of a small plane is trying to maintain a course due north. His air speed is 120 miles per hour. There is a 10 mph wind from the east. Estimate the direction in which the plane should be pointed to accomplish this.
Section 4-1

This is another question to demonstrate the effectiveness of using components.
To maintain a course due north the pilot has to compensate for the wind by giving himself a 10 mph component to the east. For that he needs an angle of $\cos^{-1}(10/120)$ north of east.

PUZZLES

"The Long Shot"

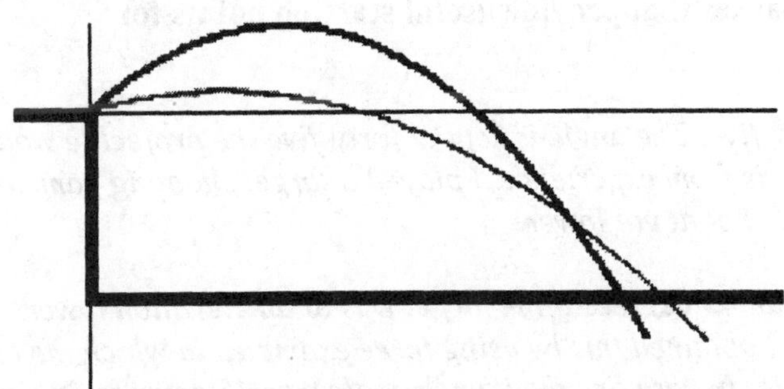

You probably know by now- from reading the book and from working the projectile motion problems- that the maximum range for a projectile is achieved when the projectile is fired at 45 degrees. This is true if the launch starts and ends at the same altitude. What about if you fire at a target at a lower elevation? Is the optimal angle still 45 degrees? Is it more? Less? Please answer this in words, not equations, briefly explaining how you obtained your answer.
Section 4-4

We usually start the puzzle discussion by displaying the most correct and complete student response, e.g., response number 4 below. The response is based on an analysis of the trajectories, so the classroom discussion would start there. With the help of the author of the response, we would construct the trajectories shown in the figure.

We would then carefully reproduce the student's reasoning and his/her conclusion that for a target at a lower elevation, a reduced angle would give a longer range. Next we would develop the "range" formula for this case. This is a nice opportunity to review the kinematics equations. Analyzing the expression, the students will agree that to obtain an analytical answer to the question, we need to take the derivative of the range expression with respect to the launch angle, set the resulting expression to zero, and solve the resulting equation for the launch angle. Two facts emerge:

1. The resulting algebra is pretty ugly.
2. The solution depends on the difference in elevation.

Now we have three options:

1. Leave it at that.
2. Write down the solution and test the limit of zero height difference.
3. Invite the students to do the algebra for extra credit.

Which option we take will depend on the class. If only one or two students are capable of option 3, it is probably best to take option 2. Notice that the student who submitted response 4 is likely to take option 3, given some incentive and probably some out-of-class help. Puzzles are usually extendible. We encourage the students to think of these extensions, and we bring them up during the classroom discussion of the puzzles. The puzzle presented here has several obvious extensions:

1. Analyze the problem for a target at a higher elevation.
2. Analyze the problem of shooting up an incline (this requires a redefinition of "range," which most students don't find obvious).
3. Analyze the problem if air-resistance cannot be neglected. This is an opportunity to bring in computer modeling.

Here are some sample (unedited) student responses that provide useful starting points for discussion:

a. *I believe that the optimal angle is below forty-five. The angle is kept at forty-five the projectile would over shot the target. The reason I believe this is from experience. I played a target shooting game on Nintendo. As the target got lower, the angle I shot at got lower.*

b. *The optimal angle would be slightly lower than 45 degrees if the target was at an elevation lower than the starting elevation for the projectile. I obtained this by using three examples in which the only variable was the launch angle of the projectile. In case one the launch angle was 45 degrees. In case two, I adjusted the launch angle to 48 degrees and found that the resultant range was lower than at 45 degrees. In case three, I used a launch angle of 43 degrees and found the resultant range to be greater than the range at 45 degrees.*

c. *No, it would be less than 45 degrees, because this will give the projectile a larger horizontal velocity vector, and less vertical velocity vector (when compared to the 45 degree shot). Since it has more time to fall (lower elevation) this will cancel out the fact that it is not going as high, and will allow the projectile to go farther.*

d. *If I fire at a target at a lower elevation, the optimal angle is less than 45 degrees. Because I want to hit the lower target, my parabolic path of trajectory will have to be lowered. In other words, the y-component for the position for the lower-target trajectory needs to be smaller than the y-component for position for the 45degree trajectory. In order to accomplish this, the angle between the ground and the cannon has to be decreased, and therefore, the angle of the projectile would be less the 45 degrees. Put more simply, the path of trajectory has a better chance at hitting the lower target at lower heights (and the smaller angle between the cannon and the ground is the way to achieve this). (To hit at a higher target, the angle would be more than 45 degrees, because the path of trajectory would be higher and the projectile would have a chance to hit the higher target).*

e. *Yes, a 45 degree launch still yields a maximum range. Since the landing spot is lower then the launch spot you need to first find the time the projectile is in the air. Next you use that time to find the distance (x) it travels. Since the final equations involve both sin and cos, to get the maximum distance, the sin and cos of an angle must be as close to 1 as possible. 45 degrees turns out to be the best angle.*

Chapter 5: Newton's Laws of Motion

WARM-UPS

1. The word "push" is a reasonably good synonym for the word FORCE.
Is "hold" a synonym for force?
How about "support"?
How many others can you think of?
Section 5-1

This vague question is intended to generate a feel for the meaning of the term "force." Most physics terms with counterparts in everyday language carry connotations that mask the essential physics. A force will cause an acceleration unless it is balanced out by other forces. Students always pleasantly surprise us with a rich collection of choices (including bad ones, such as "power").

Here are some sample (unedited) student responses that provide useful starting points for discussion:

a. Hold and support are good words because they involve interaction between objects or the environment. Other good words are pull, contact and gravity.

b. Push or pull are both good synonyms for the word force. Hold is not a good synonym because we need both a magnitude and a direction when describing a force. Support could be used if you are holding a weight then you are exerting an upward force on the object.

c. All of them describe force, as well as shove, carry, swing, hit, kick and anything that has one body moving against another.

d. I would use words more closely synonymous with push, such as shove or pull. Hold and support both are assimilated with keeping something at rest, which would give an object zero acceleration and therefore zero force.

e. No, a hold is not a synonym for force. Support is not, either. Both of those words deal with keeping an object stationary/at rest. I would not use those words to mean force. Some words I might use, would be pull, stress, power, weight, friction, tension, might, strength, intensity, energy, influence, or pressure.

f. I don't think that 'hold' is a good synonym for force. F=ma; if you are holding something then it has zero acceleration and therefore no force.

g. 'Hold' could be a force that counteracts another such that the net force is zero. The same could be said about 'support'. Pull, lift.

2. The engine on a fighter airplane can exert a force of 105,840 N (24,000 pounds). The takeoff mass of the plane is 16,875 kg. (It weighs 37,500 pounds.) If you mounted this aircraft engine on your car, what acceleration would you get? (Please use metric units. The data in pounds are given for comparison. Use a reasonable estimate for the mass of your car. A kilogram mass weighs 2.2 pounds.)
Section 5-3

Good to discuss mass vs. weight, kilograms, newtons, pounds. Also good to get a feel for the size of these units. A 1000-kg car acted upon by a 100,000N force would accelerate at about 10 g's.

3. A 72-kg skydiver is descending on a parachute. His speed is still increasing at 1.2 m/s^2. What are the magnitude and direction of the force of the parachute harness on the diver? What are the magnitude and direction of the net force on the diver?
Section 5-3

This question tries to elicit the students' grasp of Newton's second and third laws. The responses, in our experience, are usually very confused. Having given the students a chance to think about a very specific, easy-to-understand example, we get another chance to sort this out in class. The answer to the second question is simple, but not easy for the students (86.4 N, down). The first question gives us a chance to emphasize the value of free body diagrams. (619.2 N, up)

4. Suppose you run into a wall at 4.5 m/s (about 10 mph). Let's say the wall brings you to a complete stop in 0.5 s. Find your deceleration and estimate the force (in newtons) that the wall exerted on you during the stopping. Compare that force to your weight.
Section 5-3

This question refers back to Warm-Up 2 in Chapter 2. The deceleration is about 9 g. So, the force is about nine times the weight.

PUZZLES

"Stretching..."

A 10-kg mass (on the right) and a 5-kg mass (on the left) are suspended from pulleys as shown in the diagram:

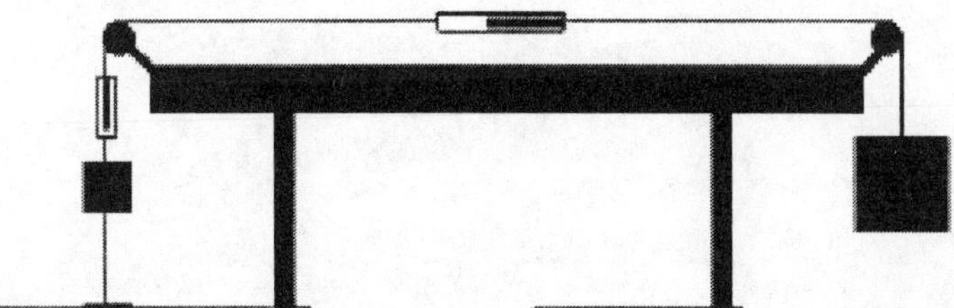

The lower mass is tied to the floor by another string. Two small metric scales (calibrated in newtons) are spliced into the rope. One of them is between the two pulley discs, the other just above the smaller mass. Assume that the masses of the scales are small enough to be neglected. What do the scales read? The string under the lighter mass is now cut. What do the scales read after the masses start to move?

A scale of negligible mass indicates the tension in the string. To determine the tensions in the strings, a student needs a free-body diagram.

Labeling the tension to the right of the horizontal scale T_R and the tension in the string above the vertical scale T_L, the free-body diagram on the right shows that $T_R = (10).(9.8)$. $T_L = T_R$
When the string is cut, things are more complicated. $T_L - 5g = 5a$ and $T_R - 10g = -10\,a$.
Since the mass of the horizontal scale is negligible, $T_L = T_R$.

So $a = 5g/15$ *and* $T_L = T_R = 5\ 1/3\ g$.

Here are some sample (unedited) student responses that provide useful starting points for discussion:

a. *While the string from the small mass is still attached to the floor, the scale between the pulleys will read 49 N and the other scale will read -49 N. After the string is cut, due to an acceleration in the direction of the larger mass, the first scale will read 74.5 N and the second scale will read 74.5 N.*

b. *Since the strings are not cut the scales will have the same reading. This is because the tension in the rope is the same. The sum of the forces in the y direction on the 5kg weight is T=mg+T2. The sum of the forces on the 10kg weight is t=mg. The tension in the rope is 98N before the rope is cut. When the string under the lighter weight is cut, the tension in the rope changes. The tension is still the same throughout the rope, but now the smaller weight is counterbalancing the larger weight. The scales should read 1/2 of the original value, which is 48N.*

c. *The scale between the two pulley discs would read zero as long as the string was not cut. This is because the forces on either side are equal. The scale just above the smaller mass would read approximately 20N (10kg)because of the tension resulting from the weight being tied down. If, however, the string is cut, both scales would read approximately 12N due to the effects of the smaller mass.*

d. *At first, both scales read 147N because the summation of both forces gives the total force. After the string is cut, both scales read 16.35N. This is because F = ma and the acceleration is 3.27m/s^2. so F = m1a1 + m2a2 = 16.35N*

e. *Both of the scales will initially read 49 N since the system is in equilibrium the sum of the forces is 0 so the tension will be the mass of the block times gravity. This applies for both scales since the smaller scale is attached to the floor. Once the string is cut both of the scales will go to 0 since there will be no tension.*

Chapter 6: Applications of Newton's Laws

WARM-UPS

1. People could not walk without friction. What friction are we talking about here kinetic friction or static friction?
Section 6-1

This is an easy question calling attention to the difference between the two coefficients of friction. The static coefficient is primarily at work in the process of walking. The foot put forward in a stride must be prevented from sliding. The same is true for rolling without slipping.

2. The coefficient of friction between a "safe" walking surface and your shoes is supposed to be about 0.6. Estimate the maximum acceleration you could attain under these conditions.
Section 6-1

As the leg muscles push on the body to produce an acceleration, the sole of the foot on the ground must be held at rest by the force of static friction. On level ground, the normal force, to good approximation, is the person's weight mg. Thus the maximum acceleration attainable can be estimated from the relation $\mu mg = ma$. *With a* μ *of 0.6, the maximum acceleration before slipping would occur can be estimated at 0.6 g or about 6* m/s^2.

3. List all the forces acting on a car negotiating a turn at constant speed on a banked road. What should be the magnitude of the vector sum of all the forces acting on this car?
Sections 5-7 & 6-5

This question asks for a free body diagram. There are two objects exerting a force on the car: the Earth and the road. For analyzing the motion of the car, it is convenient to decompose the force of the road into components perpendicular and parallel to the road surface. With proper banking, the component parallel to the road surface (the force of friction) can be reduced to zero. To negotiate a horizontal turn, the car needs a net force in the horizontal direction whose magnitude is equal to mv^2/r *where ma is the mass of the car, v is the speed, and r is the radius of curvature of the turn.*

4. Estimate the coefficient of friction between skis and snow that would allow a skier to move down a 30 degree slope with constant speed if other factors were neglected. Is it reasonable to neglect other factors, such as air drag on the skier?
Section 6-1

Another free-body diagram question. The net force on the skier must be zero (the motion is with constant speed along a straight line). The component mg sin30 of the weight along the slope must be balanced by the force of friction $\mu mg \cos 30$. *Thus we estimate* μ *to be about 0.6. As long as air drag is negligible (low speed), this is a reasonable estimate.*

PUZZLES

"Double the Fun?"

Experimenting with his toy rockets, Danny fires his model rockets three different ways. First he places a model on the ground, free of obstructions, and then he fires. Next he braces an identical model against a stiff wall and fires. Third, he positions two identical models on the grounds with their tails butting and fires them simultaneously. How do the rocket accelerations in the three experiments compare?
Section 6-4

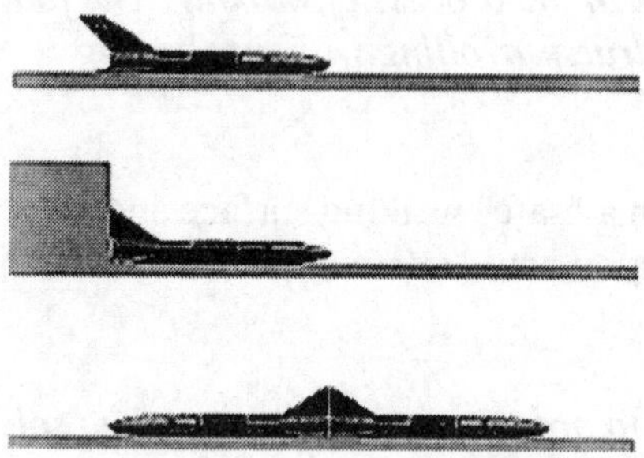

This puzzle question is trying to flush out the perception, frequently present even after classroom instruction on momentum conservation and rocket propulsion, that rockets need an external medium to "push against."

The rocket thrust is given by the (dm/dt)$v_{exhaust}$, and hence it involves only the exhaust speed of the escaping combustion gases and the rate of combustion, provided the exhaust orifice is not blocked or obstructed.

With electronic data-taking equipment such as Pasco hardware and Data Studio, the result can be checked as a classroom demo. Such an experimental check is often more convincing than a theoretical argument, particularly with a deeply ingrained preconception.

Chapter 7: Work and Kinetic Energy

WARM-UPS

1. A weight lifter picks up a barbell and
lifts it chest high,
holds it for 5 minutes,
puts it down.

Rank the amounts of work W the weight lifter performs during these three operations. Label the quantities as W1, W2, and W3. Justify your ranking order.
Section 7-1

This appears to be an easy question. The responses below suggest that, to the students, it is far from obvious.

Assuming constant speed during the lift and the drop, the ranking ought to be W1>W3>W2. W2 is zero and W3 is negative with the same magnitude as W1.

If the speed is not constant the ranking still holds, but now the magnitudes of W1 and W3 are not equal. This is a good place to introduce the work-energy theorem on a real-world example.

Here are some sample (unedited) student responses that provide useful starting points for discussion:

a. *When he is lifting the barbell would require the most work. This is because there is a force pulling it down that he must overcome. When he is putting it down is the second, because he has to fight the force pulling it down, or he would drop it on his toes. But, he doesn't have to fight it as much as when he was lifting it. The third is when he is holding it. There is no acceleration, so the force is not as great.*

b. *W1 the lifter does work on the barbell against gravity, W2 the lifter does work on the barbell but the barbell does work on the lifter also, against gravity. W3 the lifter does work on the barbell with gravity.*

c. *Work is equal to the magnitude of the weight lifted times the distance that it is lifted. Therefore since the distance and the magnitude are the same for weight 1 and 3 the work done should be the same. No work is done when the weight is being held because it is not being performed over a distance.*

d. *The most work is done lifting the barbell against gravity. When the barbell is put back down less force is exerted over the same distance, and thus less work. This is because this time the weight lifter is working with gravity, which pulls the barbell down. The weight lifter exerts a certain amount of force in order to prevent the barbell from being in 'free-fall'. The least work is done when the barbell is held up as the distance over which the weight lifer exerts the force is zero, and thus so is the work.*

e. *W2 is the greatest because he is holding the weight against gravity for a long time. W1 is the next amount of work because he is pulling to his waist against gravity. Finally, W3 is the least because he is dropping the weight with gravity.*

f. *W2 would require no amount of work on the barbell, however it would require a great amount of work within the muscles of the weight lifter. In reference to work performed on the barbell, this would be the smallest (actually zero). W1 would require the most amount of work. Since the force the weight lifter applies has to counteract the force of gravity, whereas with W3 the net force would be the sum of gravity and any force he applies (he probably would not push the weight down).*

2. Estimate the amount of work the engine performed on a 1200-kg car as it accelerated at 1.2 m/s^2 over a 150-m distance.
Section 7-1

This is a straightforward application of the work-energy theorem. Students who preview the chapter typically have no problem with this one. The answer is 864 J.

3. A real world roller coaster released at point A and coasting without external power would traverse a track somewhat like the figure below. Friction is not negligible in the real world. Does the roller coaster have the same energy at points B and C? Is the total energy conserved during the coaster ride? Can you account for all the energy at any point on the track?
Section 8-1

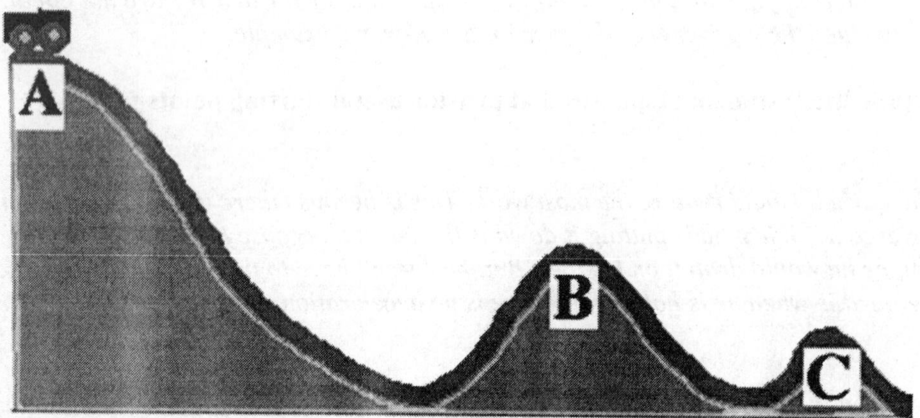

The coaster has less potential energy at C than at B. In absence of friction the kinetic energy at C would be greater than the kinetic energy at B, with the sum of the kinetic and potential energies remaining unchanged. Since friction was not to be neglected, the sum of the kinetic and potential energies at C would be less that at B by an amount equal to the work done against friction. The total mechanical energy is steadily diminishing along the track as the energy dissipative work against friction takes energy away from the roller coaster. In class, this example can be extended by actual calculations of the energy balance on straight inclines.

Here are some sample (unedited) student responses that provide useful starting points for discussion:

a. *The sum of the Kinetic and Potential energies at point B is equal to the potential energy at point A minus the energy converted to heat due to friction. The same goes for the kinetic and potential energies at C. Point C will have the lowest potential energy and point A the most.*

b. *The potential energy is greater at point B than C. Potential energy is greatest at the highest points. Kinetic energy is greatest on the steepest slope. B would have the greatest kinetic.*

c. *The potential energies at points B and C are about 50% and 25% of that at point A, respectively. Due to friction, their kinetic energies will be much smaller; that's why the hills on roller coasters are successively smaller.*

d. *The potential and kinetic energies at points B and C will be proportional to the height at those points. Since there is kinetic energy lost to friction as the coaster rolls up the hill, potential energy will be equal to the difference between the kinetic energy as the coaster leaves the top of the first hill and the kinetic energy as the coaster reaches the top of the second hill. Since the forces ofgravity and friction are constant, the decrease in kinetic and potential energies will be seen as constant as the coaster makes its up and down run.*

4. A gallon of gasoline contains about 1.3×10^8 joules of energy. A 2000kg car traveling at 20 m/s skids to a stop. Estimate how much gasoline it will take to bring the car back to the original speed. To complicate matters further, consider the fact that only about 15% of the energy extracted from gasoline actually propels the car. The rest gets exhausted as heat and unburned fuel.
Section 8-2

This question extends the energy idea beyond mechanics and attempts to make a connection between the abstract notions of the chapter and the real world. Stopping, the car loses 400,000 joules of kinetic energy. This energy has to be restored when the car is brought back to speed. In the car engine, a gallon of gas yields 19,500,000 joules of energy. The car can get the kinetic energy back by burning 0.02 gallons of gas.

Here are some sample (unedited) student responses that provide useful starting points for discussion:

a. *I'm not sure if we can answer this with the information. Wouldn't we have to know how fast it accelerates, because it will use more gas if you step on it, but if you let it idle and do it gradually, it will take less. I'll be interested to see the solution in class.*

b. *Since only 15% of the energy of the gasoline is used by the car, the vehicle gets 19.5 MJ of energy/gallon. The energy required to get the car to 20m/s is equal to half its mass times the square of its Vf. This means that the car needs 400KJ of energy to reach 20m/s. This shows that the car needs .021 gallons of gasoline.*

c. *By the work energy theorem, it would take 400,000J of energy to get the car up to speed again. So if we multiply by the ratio of gallons to energy that we were given, we get .00308 gallons are needed to propel the car. However, if only 15% of the gasoline is helping us, we need to divide to amount of gas by .15 to give us a final answer of .02 gallons.*

d. *I estimated that it would take 4 x 10^8 J to get the car back to its original speed, which is about 3 gallons of gas, and if only 15% propels the car, it would take 7 gallons. This doesn't sound very accurate to me. Considering my car this means that it would take me 7 gallons of gas to go .02 mi. I don't think I could afford that! I think that you should explain the second problem in class because I thought that I knew what I was doing but my answer doesn't seem realistic.*

PUZZLES

"Who Is Right?"

Bill is riding in a railroad car. He throws a ball toward the car wall. The train is moving at constant 20 m/s to the right. The ball is flying away from Bill at 8 m/s. According to Bill, his 0.145-kg baseball has 4.64 joules of kinetic energy. His brother is standing on the ground disagreeing. According to him, the baseball has 10.44 joules of kinetic energy. Which brother is right? Please answer this in words, not equations, briefly explaining how you obtained your answer.
Section 7-2

Both are correct, each in his own inertial coordinate system. The work-energy theorem and energy (and momentum) conservation laws hold equally well in any inertial reference frame. The work to stop the ball in the reference frame of the moving railroad car would be 4.64 J, just as Bill claims. It is possible to throw the ball so that it has no kinetic energy relative to the ground frame of reference, and the ball would still cause damage when it hit the wall of the car. In a more advanced class this might be a good place to talk a little about laboratory vs. center-of-mass coordinate systems.

Here are some sample (unedited) student responses that provide useful starting points for discussion:

a. *Bill is right because kinetic energy only depends on the baseball's mass and speed not on direction of motion. The baseball is throwing at 8m/s whether it is right, left does not matter and the fact the ball weighs 0.145kg. K.E.= 0.5(0.145kg)(8m/s)^2=4.164J*

b. *They each are right according to their own inertial frames of reference. When the frame of reference is the railroad car, the velocity is 8 and using 1/2mv^2 it works out to be 4.64. When the earth is the frame of reference the ball is moving 12m/s and it has 10.44 J of kinetic energy.*

c. *Both are correct. Kinetic energy is a relative measurement, since one of its components (velocity) is relative to the observer. For example: You need a point of reference to determine the velocity of the ball. If you choose Bill as the point of reference, the velocity of the ball is -8 m/s (choosing right as +), whereas choosing his brother as the point of reference would give the ball a velocity of 12m/s (the initial 20 minus the 8 that Bill forces it to move). If we were to choose Alpha Centauri as the point of reference, the net velocity of the ball would be quite complex to calculate, but it would probably be very high.*

d. *Bill is right. Since Bill is riding on the train he is going the same velocity as the train, therefore he only believes that the ball has 4.64j of energy. When he releases the ball the friction that was holding him and the ball where they were stops. The ball speeds toward the train while the train speeds toward the ball at 12m/s giving the illusion to his brother that the ball has a KE of 10.4J. The ball itself actually only has KE of 4.64J.*

e. *Assume that the Earth is a suitable inertial frame of reference (neglecting the motions associated with its rotation and its movement around the sun). In this case Bill's brother would have the correct answer of 10.44 joules because Bill does not have a suitable inertial frame of reference. According to Bill, the ball has a velocity of -8 m/s (If you set up the coordinate system to have right be positive). However, Bill himself has a velocity of 20 m/s because he is inside the train. Bill's brother, standing on the Earth is going 0 m/s, so he can correctly calculate the net velocity of the ball. This net velocity will give you the proper kinetic energy of the ball. A simple way to visualize this is to imagine what would happen if the train suddenly disappeared while the ball was in motion. Would the kinetic energy of the ball suddenly change because the train disappeared? No! Would the ball be moving 8 m/s to the right? No, it would be moving 12 m/s to the right, giving you the proper net velocity to use in calculating the ball's kinetic energy.*

Chapter 8: Potential Energy and Conservative Forces

WARM-UPS

1. Do frictional forces always cause a loss of mechanical energy?
Section 8-4

Frictional forces lead to a loss of mechanical energy if they involve work. This happens if the object is sliding against a force of friction or moving through a viscous medium where the particles of the medium slide against the surface of the moving object. If no sliding is present, no work is done. Examples are walking or rolling without slipping.

This may be a good place to talk about the energy loss in a tire that is deforming as it rolls. Even if the point where the tire makes contact with the ground does not slide, there is still an overall loss of energy as the tire deforms as it rolls. Why don't we go to non-deforming tires, such as steel wheels on train cars?

2. A good, professional baseball pitcher throws a ball straight up in the air. Estimate how high the ball will go. (A good throw can reach 90 mph.)
Section 8-4

This is a straightforward application of conservation of energy. $\Delta K + \Delta U = W$
During the pitch there is a W provided by the thrower. During the assent there is W involving air drag. If air drag can be neglected, the ball will reach about 83 meters. As students will often answer this question using kinematics, this is an opportunity to bring out the virtues and advantages of the energy method.

3. Which potential-energy U(x) versus x graph corresponds to the force F(x) versus x graph shown?
Section 8-5

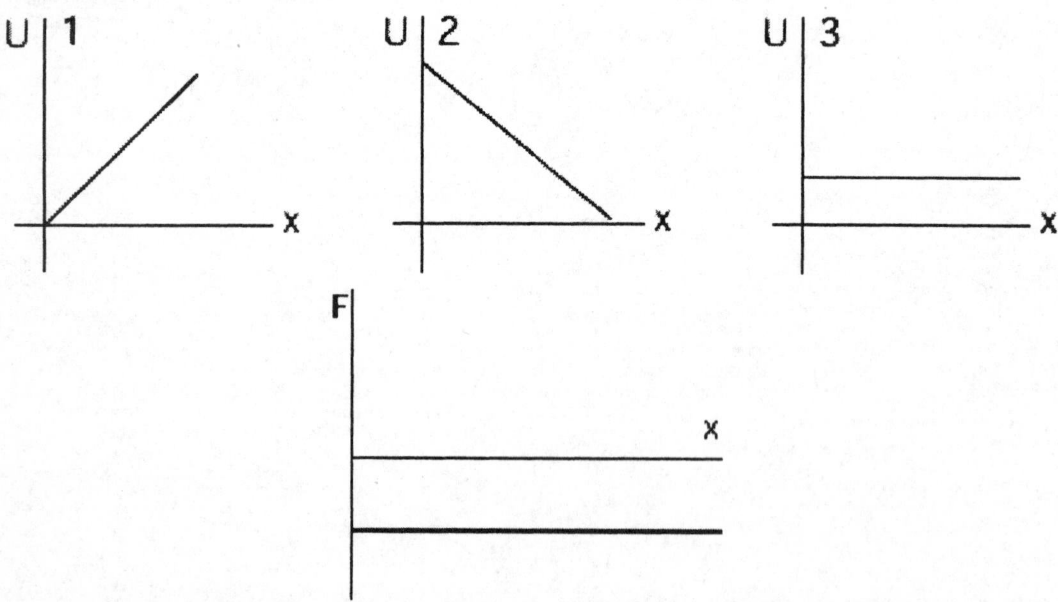

A discussion starter for potential-energy diagrams. Graph 1 represents the potential energy.

4. Estimate the energy burst you would need to clear a 1-meter hurdle.
Section 8-4

This is similar to Question 1. To clear the hurdle you need to increase your potential energy by mgh. Taking 80 kg as a reasonable value for the mass, the energy burst needed is about 800 J.

PUZZLES

"Energy Drop"

Trapeze artist Lea stands at point A. Under her knees she attaches a taut ten-meter rope, tied to the ceiling at the point marked with a dot. She lets herself go. At the bottom of the quarter circle she picks up her twin sister, who is standing at point B. Together they continue on the 10-meter radius circle until they come to a stop at the platform marked C. Between platforms B and C they rise vertically 2.5 meters. Can you account for all the energy transformations in this stunt?
Section 8-3

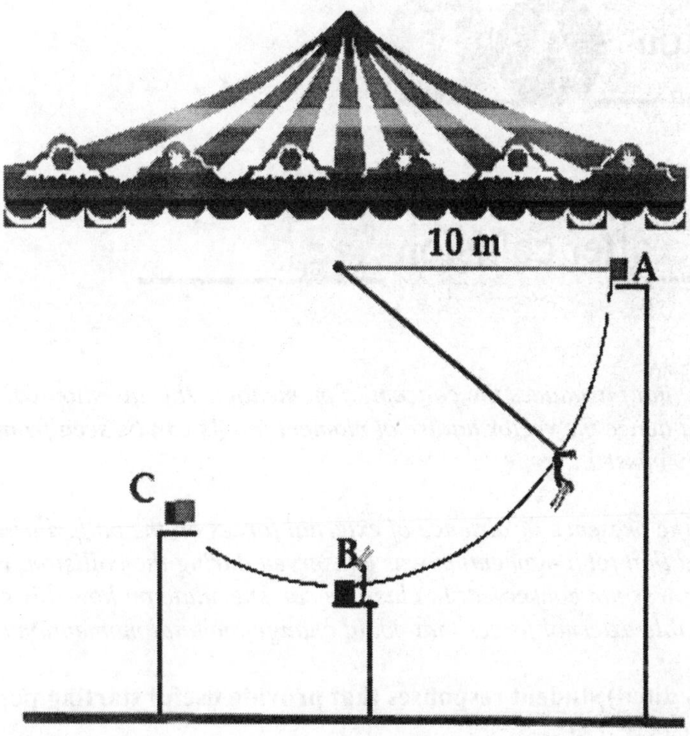

At point A, Lea has potential energy relative to point B. The potential energy is transformed into kinetic energy. When she picks up her sister they take off together. There is no reason to assume that they share the original kinetic energy that Lea brought into the interaction. At this point the student has no way of knowing how much energy is still available. Since there was an interaction at point B the student may not assume that the all the kinetic energy was transferred to the pair. In fact, it was not, as the next chapter will show. Whatever kinetic energy the pair leaves point B with, gets transformed into potential at point C.

Chapter 9: Linear Momentum and Collisions

WARM-UPS

1. Ball A in the diagram is chasing after ball B. The speed of ball A is 3 m/s, the speed of ball B is 2 m/s. After the collision ball A rebounds at 2 m/s and ball B speeds away at 3 m/s. What was the change in velocity experienced by ball A during the collision? What was the change in velocity experienced by ball B during the collision? If the mass of ball A is the same as the mass of ball B, was momentum conserved during this collision?
Section 9-1

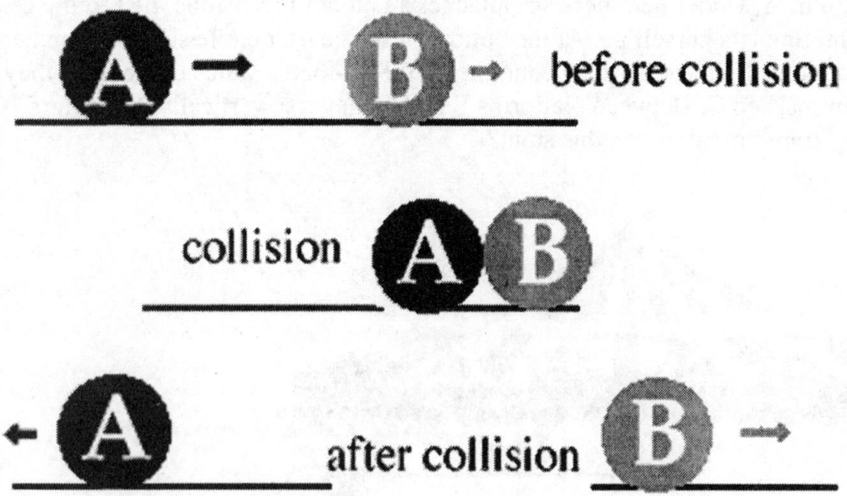

As a warm-up for the lesson that introduces the concept of momentum, this question calls attention to the vector nature of velocity and hence the vector nature of momentum. As can be seen from the student responses, this point is easily missed.

Since nothing is said about the presence or absence of external forces on the balls during collision, it cannot be concluded a priori that total momentum was conserved during the collision. From data given we conclude that momentum was not conserved. In class we can speculate on how this could have occurred. What are the possible external forces that could change the total momentum of the balls?

Here are some sample (unedited) student responses that provide useful starting points for discussion:

a. *The change in velocity of ball A is 2 m/s - 3 m/s=-1 m/s. The change in velocity of ball B is 3 m/s - 2 m/s=1 m/s. Total momentum was conserved from the collision.*

b. *The change of velocity of ball Al is -5 m/s. The change of velocity of ball B is 1 m/s. The net external forces are zero thus momentum is conserved.*

c. *No momentum was not conserved because if you use the equation p= m1v1 + m2v2 before the collision and after the collision, they are not equal like they should be.*

d. *The velocity of ball A decreased by 1m/s during the collision. The velocity of ball B increased by 1m/s during the collision. Yes momentum was conserved, because before the collision the momentum of the system is 2m + 3m = 5m, and after the collision the momentum is 3m + 2m = 5m. Thus momentum is conserved.*

2. Estimate the force on a 145-g baseball if it hits the bat at 50 miles per hour, stays in contact with the bat for 0.1 second, during which time it reverses direction and leaves the bat at 75 miles per hour. Express the result in newtons and in pounds.
Section 9-3

Again, directions must be observed carefully to arrive at the answer of 81 newtons or 18 lbs, opposite the original direction of motion of the baseball.

3. Suppose the shuttle was launched from a launch pad on the Moon. What changes would we observe? Suppose it was launched from a launch pad in space in a vacuum. Would it work?
What changes would we observe?
Section 9-8

For the same thrust (which only depends on the mass and exhaust speed of the gases, the net force on the Moon would be greater than on the Earth, and even greater yet in deep space, away from major gravitational pulls on the shuttle. So the accelerations would vary accordingly.
The question also flushes out the notion that the exhaust gases must push against "something." Students who hold that view have illustrious company

"Professor Goddard does not know the relation between action and reaction and the need to have something better than a vacuum against which to react. He seems to lack the basic knowledge ladled out daily in high schools."
(1921 New York Times editorial about Robert Goddard's revolutionary rocket work.)

Here are some sample (unedited) student responses that provide useful starting points for discussion:

a. *From the moon, the shuttle would take off at such a high velocity it would almost be ridiculous, because of the decreased force of gravity. If it were launched in the vacuum of space the results would be very amusing. The shuttle would barely move, but the launch pad would be, well, launched in the opposite direction.*

b. *If the shuttle were launched from the moon, it would need to produce less energy to escape gravity. Thus the change in its mass would be smaller compared to earth. In a vacuum there would be no air resistance to fight. This eliminates another downward force to overcome. It would take less energy to leave the earth. So less fuel is used and thus the mass would not change as much.*

c. *If the shuttle were launched from the moon as is, there would be a tremendous increase in the velocity because the force of gravity is much smaller. It could also be launched from space in a vacuum, because the thrust of the rockets would cause a force in the opposite direction, moving the shuttle.*

d. *It would take less work (J) to launch the craft because the moon has less gravity. If it were launch in a vacuum in space it would work very well because there would be practically no gravity. Also, it would take hardly any work to launch the craft.*

e. *ON MOON: It would launch a lot faster, because the force that the ejecting gases exert on the shuttle would be the same as an earth launch, but the force of gravity on the moon is less than the force of gravity on earth. The shuttle would have a much greater net force on it. Also the launch wouldn't be very loud since there is not much of a lunar atmosphere.*
IN OUTER-SPACE: There would be a super-quick launch because there is no gravitational force in opposition to the exhaust force. Even human waste ejecting from the spacecraft would affect its motion. It would work, because momentum is conserved, and if the exhaust is going really fast in one direction, then the shuttle has to be going in the opposite direction so the total momentum is conserved.

f. *We would need less force to propel the rockets on the moon. In vacuum, I don't think it would work because there would be nothing to "push on."*

4. Ball A is propelled forward and collides with ball B initially at rest.
After the collision, the balls' trajectories are as shown in the diagram.
How do the masses of the two balls compare? Briefly explain your answer.
Sections 9-5 & 9-6

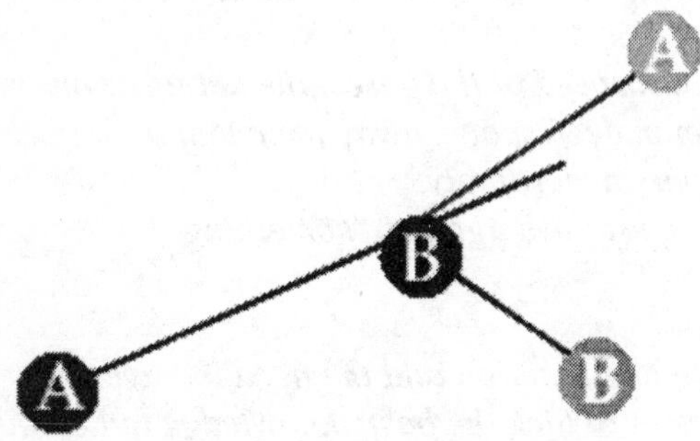

Since the total momentum in the direction perpendicular to the initial momentum must be zero, we conclude that ball A is more massive than ball B. The first student response gives a good version of this argument.

Here are some sample (unedited) student responses that provide useful starting points for discussion:

a. *If we place a coordinate system with the x-axis pointing along the initial trajectory of ball A before the collision, then there is no component of the total momentum in the direction perpendicular to the x-axis. Since momentum is conserved, the total momentum of the two balls after the collision must also be zero. Ball B goes farther than the ball A in the y-direction so ball B has a greater magnitude y-velocity. Since the product of the mass and y-velocity of both balls are equal in magnitude, but opposite in sign, then ball B has a smaller mass, since it has the greater y-velocity.*

b. *Really there is nothing conclusive we can say without some more information. Ball A may be much more massive than ball B and strike ball B slowly or ball A may be less massive but graze the higher mass of ball B with a high enough velocity to cause ball B to roll as shown. We need additional information.*

c. *The mass of ball A is greater than the mass of ball B because the A ball's final trajectory is more along the initial path because it's momentum was greater.*

PUZZLES

"Knock-Off"

You are standing on a log, and a friend is trying to knock you off. He throws the ball at you. You can catch it, or you can let it bounce off of you. Which is more likely to topple you, catching the ball or letting it bounce off? Briefly explain what physics you used to reach your conclusion.
Section 9-3

You are better off catching the ball. What topples you is the force the ball exerts on you, which is the reaction to the force you exert on the ball. The force you exert on the ball changes the momentum of the ball.

The change of momentum = $m_{ball}\, v_{final} - m_{ball} v_{initial}$.

The magnitude of v_{final} *will be greater if you bounce the ball, hence the force will be greater also.*

Chapter 10: Rotational Kinematics and Energy

WARM-UPS

1. In rewinding an audio or videotape, why does the tape wind up faster at the end than at the beginning? Section 10-2

(The following is taken from Novak et.al. Just-in-Time Teaching: Blending Active Learning with Web Technology (Prentice Hall, 1999), where further discussion of this question appears on page 51.)

Consider the questions a student faces in analyzing and modeling this situation. There are two spools that are spinning (same rate or different rate?), and the tape is winding on one while it is unwinding on the other. Does the term "faster" in the question refer to the spools, the tape, or both? The students have to try to relate the idea of "winding the tape" to one or more technical terms in the chapter. Students may notice the relationship of "winding" to "rolling without slipping." This illustrates the extendibility built into many of the Warm-Up questions.

To answer the question, students must find relationships among the quantities they have identified. They typically draw these from statements they have been told to believe, or from inferences they draw from their own experiences. For instance, many students have heard sportscasters discuss the rotation speed of ice skaters, divers, etc. The conclusion is "smaller radii imply higher rotation speeds." The eighth response above demonstrates this. They also tap technical knowledge gained from previous courses or from the textbook.

To prepare for the interactive lecture session, we collect the student responses, read them, and group them into categories. Some categories are general, and some are subject-specific. General categories include responses from students who don't know where to begin, those who misunderstand the question or deliberately answer a different question, those who correctly describe the scenario but never answer the question, and those who bluff their way through the answer. Since responses in these categories are generally not of interest to the entire class, we usually try to deal with them individually, such as through a face-to-face discussion or e-mail exchange with the student. Occasionally, however, a creative misinterpretation of the question can be the starting point of an interesting class discussion

Here are some sample (unedited) student responses that provide useful starting points for discussion:

a. *Because the gear in the rewinder is having to turn less amount of weight at the end.*

b. *Because the tape has a much larger diameter to wrap around. This causes the tape to wind faster.*

c. *It winds up faster at the end, because it has lesser and lesser mass to be rotated from the other side. Thus, as it winds up, the force from the opposite way constantly decreases. The spool that the tape is winding onto has a much smaller radius than the spool full of tape so it has to rotate several more times than the other spool to wind up the same amount of tape, once you go past equal radii of the spools and get near the end the opposite is true because the right spool is rotating the same amount but the other spool has to rotate much faster to keep up, therefore faster winding.*

d. *Because the tension is less due to less tape and therefore less resistance to turn.*

e. *The VCR will make the 'tape spools' rotate at a constant speed, given them a constant angular velocity. When you start rewinding the tape, the majority of the tape as on the 'source' spool waiting to be transferred to the 'target' spool. Initially the target spool has a small radius (due to no or little tape being on it). This means the tape will have a high linear velocity, making it rewind faster. As more tape gets transferred to the target spool, the radius of the target spool gets larger, decreasing the linear velocity, making it rewind slower.*

f. *The tape winds up faster at the end then at the beginning because it requires more tape per revolution, during the same time frame, at the end then at the beginning. As the tape narrows the end, the diameter of the side it is taking tape from becomes smaller and smaller. Since the tape is moving at constant V, the RPM must be greater to get the same amount of tape off the reel.*

g. *Just like with the skater example, as the tape pulls in closer to the rim, the rotational inertia decreases so the angular speed (tape winding) increases.*

h. *When the tape gets big on the other side the velocity is greater in turn pulling the tape faster from the small end.*

2. Estimate the magnitude of the tangential velocity of an object in your hometown, due to the rotation of the Earth.
Section 10-3

This somewhat difficult question challenges the student to think through the relationship between angular and linear quantities. They have to find the correct angular velocity for the problem ($7.27x10^{-5}$ rad/s^2), and they must find the correct radius R for the relation $v = \omega R$. ($R = R_E \cos\theta$, where R_E is the radius of the Earth and θ is the latitude of their hometown.)

3. A small object is sliding inside a circular frictionless track. At point A another frictionless track switches it to a smaller circle (see the picture). At all times the tracks can only exert forces that are perpendicular to the motion of the little slug. What happens to the linear velocity of the slug when it switches tracks? What about its angular velocity?
Section 10-3

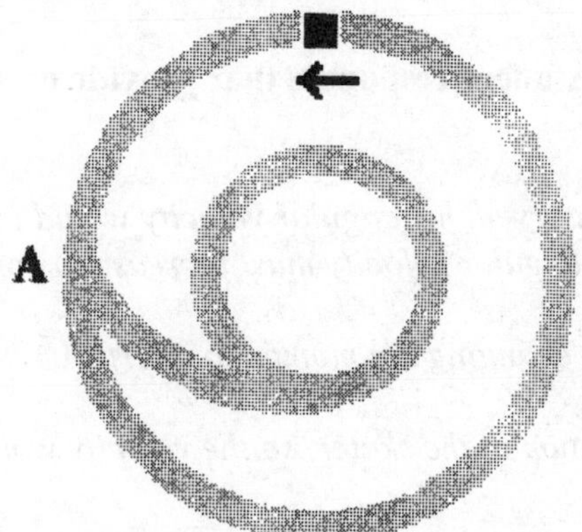

Since the external force on the slug is perpendicular to the slug's linear velocity at all times, the linear speed of the slug does not change, although the linear velocity does change because the direction is changing. As the slug moves into the track between the two circles, the force on the slug is no longer directed toward the center of the circles, and therefore there is an external torque and hence the angular

momentum must change. Finally, the angular velocity, being equal to v/r, is greater in the smaller circle (v remains the same, r is less).

Here are some sample (unedited) student responses that provide useful starting points for discussion:

a. *No, it remains constant because it's frictionless and the only force exerted on it is perpendicular to the motion.*

b. *If the track only exerts forces that are perpendicular to the slug, the slug will not speed up or slow down, it will only change direction. Angular momentum is velocity times mass times position vector R. When the slug moves from the big track to the small track its mass does not change neither does its speed. Since the value of R depends of the choice of the origin and that value stays constant to, then the angular momentum of the slug stays constant.*

c. *The angular momentum of the slug does not change. There is no net external torque on the slug. Thus angular momentum is conserved.*

d. *The angular momentum changes in a way that is equal to the torque of the net force acting upon the slug. Since there is a net force acting on the slug to move it from the outer track to the inner track, there would be a change in angular momentum.*

4. A skater is spinning with his arms outstretched. He has a 2-lb weight in each hand. In an attempt to change his angular velocity, he lets go of both weights. Does he succeed in changing his angular velocity? If yes, how does his angular velocity change?
<u>Section 10-5</u>

Letting go of the weights means that he opens his hands and releases the two masses. They fly off tangentially to the original motion and they drop since he no longer holds them against gravity. The masses carry off their original angular momentum, but this does not change the state of motion of the skater. No external torque is introduced when he lets go of the weights. Hence, his angular velocity does not change. In class, compare this action to the case where he holds on to the masses but stretches out his arms or pulls them in. Compare also the case where the spinning skater picks up two stationary masses.

Here are some sample (unedited) student responses that provide useful starting points for discussion:

a. *Since angular momentum is conserved, his angular velocity would increase after he dropped the weights. M was reduced, so his angular velocity must increase to conserve angular momentum.*

b. *It increases. This is the same as changing the moment of inertia.*

c. *Yes, the mass effects the momentum of the skater, so the velocity would change according to the change in mass.*

d. *Yes. The weights carry all the momentum. When he lets go of the weights, his momentum drops to zero. Thus his velocity also drops to zero.*

PUZZLES

"Lookalikes"

You are given two identical-looking metal cylinders and a long rope. The cylinders have the same size and shape and they weigh the same. You are told that one of them is hollow, the other is solid. How would you determine which is which using only the rope and the two cylinders?
Section 10-5

Since the masses and the shapes of the cylinders are the same, the hollow cylinder will have a larger moment of inertia. More mass is distributed away from the axis of the hollow cylinder. Thus a rotation test might distinguish between the two moments of inertia. A yo-yo-like arrangement will make the solid cylinder roll down the rope with greater acceleration than the hollow one.

Here are two sample (unedited) student responses that provide useful starting points for discussion:

a. *You could make a pendulum tying the rope to each cylinder one a time and swinging it time how long it took to stop. The one that took more time to stop is heavier. I would determine this by tying each cylinder to the rope, rolling them up and letting the rope unwind. By the time that they each reach the end of the rope, they will have the same loss of potential energy. All of this will be transferred into total Kinetic energy, which is rotational kinetic energy plus linear kinetic energy. The rotational kinetic energy for the hollow one will be large since its moment of inertia is large. Therefore, its linear kinetic energy will be a lot smaller. It will take a lot longer to unwind. In the same respect, the solid cylinder will have a small rotational K and a large linear K, this will mean that it will be going much faster since their masses are the same.*

b. *Basic idea: The two cylinders have different moments of inertia. I(solid) < I(hollow) ('University physics', p.278) You wind one end of the rope round one cylinder, and the other end around the second cylinder, until they meet in the middle of the rope. Then you let both cylinders fall freely from 4-5 feet while holding the rope in the middle (you release them at the same time). The rope starts unwinding, and you notice that one side unwinds faster. Explanation: The only forces acting on the cylinders are their weights (the gravitational force) and are therefore equal. This means that the total work done by gravity is the same for both cylinders if they fall from the same height. From the demo in class I know that the hollow one would roll down an incline plane slower than the solid. Using the rope to simulate a yo-yo, the solid would travel down and back up the rope more quickly than the hollow cylinder. The hollow cylinder has a much higher Moment of Inertia than the solid cylinder. If you tie one cylinder to each end of the rope and spin it, the hollow disk will be far more difficult to get spinning than the solid one.*

Chapter 11: Rotational Dynamics and Static Equilibrium

WARM-UPS

1. Two solid discs are linked with a belt. The diameter of the larger disc is twice as large as the diameter of the small disc. If the small disc is rotating at 500 revolutions per minute (rpm), what is the rotational speed of the large disc? Express your answer in rpm and in radians.
Section 10-3

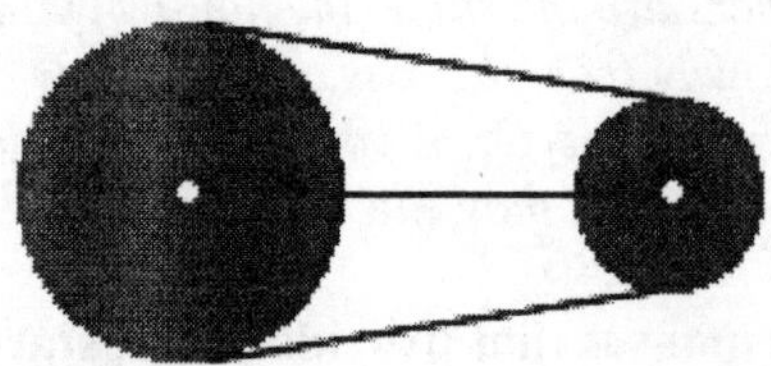

The student is asked to review the relations between linear and angular quantities. The coupling of the two discs is such that they have a common tangential velocity. Thus the ratio of their angular velocities is the inverse of the ratio of the radii since $v = \omega r$ and v is common. So the angular speed of the large disc is 250 rpm.

How would the discs have to be coupled so that they would have a common ω? *This question can lead to the discussion of more complicated systems, e.g., a hanging weight passing over a pulley, pulling a wheel that roles without slipping.*

2. Estimate the angular acceleration of a small pebble stuck to a bicycle tire as the bicycle accelerates from rest to 10 mph (4.47 m/s) in 2 seconds.
Section 10-1

Another review question from kinematics in preparation for Newton's Law for rotation. The student has to make the connection between the acceleration of the bicycle wheel (center-of-mass) and the tangential acceleration. Students also have to realize that they need the radius of the tire to answer the question. With an estimate of the wheel radius of 0.5 m, the answer is 4.47 rad/s^2. Discussion question: How does the bike speedometer work?

3. A book can be rotated about many different axes. The moment of inertia of the book will depend upon the axis chosen. Rank the choices A to C above in order of increasing moments of inertia.
Section 10-5

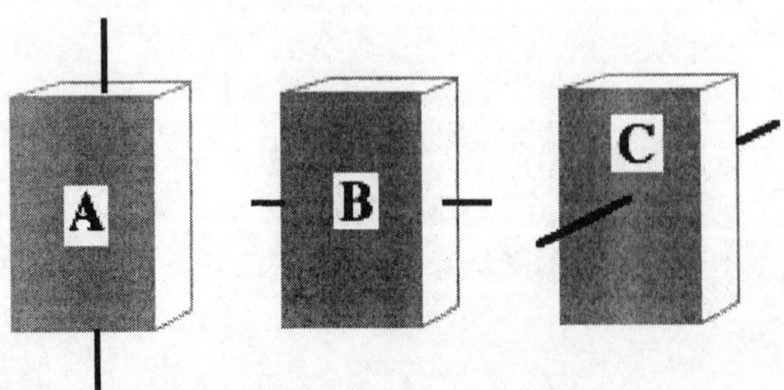

This is intended to be a conceptual question, reasoned out from basic principles, not a calculus exercise. Clearly in setup A, the mass of the book is closest to the axis of rotation so the moment of inertia of choice A is the smallest. The farthest from the axes are the corners of the book in choice C. So C would have the largest moment of inertia. Choice B is in-between. Note: After having reasoned it out, check Table 10-1.

4. A hoop and a solid disc of the same mass and the same radius are released from rest at the top of an incline and allowed to roll down the incline without slipping. Which are correct?

a. The hoop has a larger moment of inertia than the disc
b. Gravity is the only force that exerts a torque on the hoop and the disc.
c. Both gravity and friction exert a torque on the hoop and the disc.
d. Gravity exerts a greater torque on the hoop than on the disc.
e. Gravity exerts a greater torque on the disc than on the hoop.
f. Gravity torque on the hoop has the same magnitude as gravity torque on the disc.
g. If the net torque on the hoop has the same magnitude as the net torque on the disc, the angular acceleration of the hoop is the same as the angular acceleration of the disc.
h. If the net torque on the hoop has the same magnitude as the net torque on the disc, the angular acceleration of the hoop is smaller then the angular acceleration of the disc.
i. If the angular acceleration is smaller, the tangential acceleration is smaller.
j. If the tangential acceleration is smaller the center-of-mass acceleration is smaller.
k. The object with smaller tangential acceleration will take longer time to reach the bottom of the incline.

Section 10-6

This scenario appears in most textbooks and is often done as a classroom demonstration. This question prepares the student for the formal analysis. The student has to think about the problem in small, manageable steps. Note that in this question the two objects have the same mass and the same radius. In the formal analysis in class it is of course important to point out that even a small diameter hoop will lag behind a large diameter disc. Ask the students why the radius drops out of the equations. The true statements in the above set are: a, c, f (about what point?), h, i, j (if they roll without slipping), and k.

PUZZLES

"Dizzy Ball"

You practice throwing baskets from a spot 15 feet from the pole until you can make a basket every time. The ball leaves your hands the same way every time. The initial velocity of the ball is exactly correct. You visit an amusement park where there is a merry-go-round with a 15-foot-diameter platform. A standard size basketball post is fastened to the rim of the platform. The platform is rotating clockwise (as seen from above) with a period of 15 seconds. Step onto a spot at the edge of the platform, diametrically opposite the basketball post. Imagine a ground-based coordinate system, with you at the origin, the positive x-axis drawn to your right, the y-axis drawn away from you toward the post and the z-axis drawn straight up from the spot you are standing on. When the platform is standing still you can always make the basket with the v_y and v_z components of the initial velocity you practiced on the ground. How would you have to adjust the initial velocity of the ball to make a basket on the moving platform? Do you need to add an x-component to the velocity? Do you have to adjust the y-component and the z-component in any way?

Section 11-9

This is intended to be a qualitative reasoning question. We do not expect students to come up with numerical values for the velocity components that will land the ball in the basket. The rotating platform is not an inertial frame of reference. All the components of the ball velocity have to be adjusted. When the ball leaves the hand, it will follow a parabola in space, determined by the initial velocity. The initial velocity in the z-direction is determined by the throw, the initial velocities in the x- and ydirections are determined by the throw and the motion of the platform. The kinematics equations determine the subsequent position (x,y,z) of the ball in the ground frame of reference. The basket moves along a circle at constant z, with x and y-coordinates determined by the rotational speed of the platform. To hit the basket, the ball has to be given the correct velocity vector so that the position of the ball coincides with the position of the basket when the z-coordinate of the ball coincides with the z-coordinate of the basket on the way down.

Chapter 12: Gravity

WARM-UPS

1. A planet orbits the Sun along an ellipse. Consider points A and B on the ellipse. How does the centripetal force exerted on the planet at point A compare with the centripetal force exerted on the planet at point B? How about the potential energies at A and B? Kinetic energies?
<u>Sections 12-1 & 12-3</u>

Getting ready to discuss Kepler's laws. The centripetal force is provided by the gravitational attraction of the central body. It is greater at A than at B. Neither force is correct for the planet to maintain its distance from the Sun (i.e., a circular orbit.) This is a good time to ask the students what it would take for the planet at either A or B to maintain the distance, i.e., prompt them to derive the speed equation for circular orbit.

The potential energy at A is less than the potential energy at B (with what convenient zero for P.E?), the kinetic energy is larger at A than at B (why is total energy conserved?)

Note that the student answers below show some confusion regarding the magnitude and the algebraic sign of potential energy. This may be a good time to address that issue.

Here are some sample (unedited) student responses that provide useful starting points for discussion (Note that most students realize that the centripetal force is the gravitational attraction of the Sun and do not get confused by the shape of the orbit):

a. *The planet's velocity varies so that the ratio of area to time is constant, regardless of the planet's position in its orbit. And since centripetal force, potential energy, kinetic energy and angular momenta are all dependent on velocity, then they too are constant.*

b. *The centripetal force at point A would be greater than the centripetal force at point B because the radius is greater to point B. The potential energy at point A would be less than that at point B because of the same reasoning with negative signs. The kinetic energies and angular momentums at both points would be equal because the mass and velocities stay the same.*

c. *The centripetal force of A is greater than B. A is closer to the sun, which is the planet's source of gravitational attraction. Since PE is inversely proportional to the distance between the planet and the sun, A will have a very large PE compared to B, which will have a very small PE.*

d. *In an elliptical orbit, angular momentum is conserved. If this is true, then the velocity must be greater at point B resulting in a higher kinetic energy. The centripetal acceleration is greater at point B also because of the greater velocity. The potential energy at point B is greater because it is further away.*

e. *The centripetal force at A is greater than at B because the centripetal force at both of these points is simply the magnitude of the gravitational force, but the gravitational force at A is greater. This is because the planet is closer to the sun.*

f. *The potential energy at B is greater than at A, because it is farther away, and in coming from B to A, gravity would do work on the planet which decreases its gravitational potential energy. Energy is conserved, so as the potential decreases, the kinetic increases. Thus the kinetic energy is greater at A.*

2. Estimate the force that the Moon exerts on you when it is directly overhead.
(You need some data from the book to answer this question.)
Section 12-1

This is a simple application of the gravitation equation. Using 100 kg as the default human-mass estimate, the attraction of the Moon is about 3×10^{-3} N, about a million times less than the attraction of the Earth.

3. Two different planets are orbiting the same sun along two different orbits. The inner orbit is circular, the outer orbit is elliptical. Compare the speeds of the planet in the outer orbit at points A, B, and E. Compare the speeds of the planet in the inner orbit at points C, D and E. The planet in the outer orbit is to be shifted to the inner orbit as passes point E. Does it have to speed up or slow down?
Section 12-3

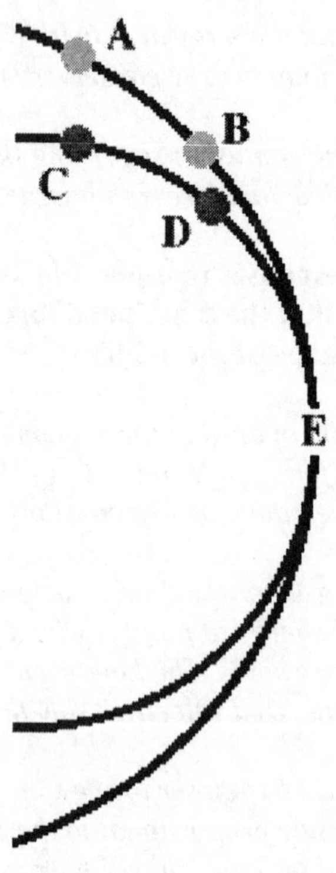

This is an elaboration on Question 1. Note that for the elliptical orbit, the point E is the perihelion. (It is worth asking the student to justify this. Since both planets orbit the same Sun, the Sun must be at the center of the circle and thus E is the perihelion for the ellipse.) Since point E is the perihelion, the planet's speed in the elliptical orbit is increasing as it travels from A to B to E, $v_A < v_B < v_E$.

In the circular orbit the speed is the same at all points so $v_C = v_B$.
When the planet reaches point E on the elliptical orbit, its speed is too fast, and therefore its distance from the Sun starts to increase. To maintain the distance (i.e., go into the circular orbit), the planet would have to slow down. (Review the formula for the speed in a circular orbit.)

Here are some sample (unedited) student responses that provide useful starting points for discussion:

a. *Speeds are : E > B > A, they speed up as they go around the curve. The speeds at C, D, and E are the same since it's a circle, every spot is the same. It would have to slow down, a tight curve is quicker than an even circle.*

b. *The orbits are moving with a constant velocity thus angular momentum is constant. In the elliptical orbit at any given point in its orbit its sector velocity is the same value I also think that the points C and D also have the same value since velocity is constant. Also, I think that the circular orbit is moving faster than the elliptical orbit because the value for R in the elliptical orbit is larger than that of the circular. Also, at point E neither orbit has to speed up or slow down, since the circular orbit is moving at a faster velocity point C will reach point E first and then point A will reach point E.*

c. *A planet travelling in an elliptical orbit travels with a greater velocity at shorter distances from the sun which it orbits, therefore, the velocities of the planet in the elliptical orbit can be ranked as follows: E > B > A. A planet in a circular orbit would travel at a constant velocity, therefore, the velocities of the planet in the circular orbit are equal at C, D, and E. At point E the planets have equal velocities, so to stay in the elliptical orbit, it would have to slow down but to switch to the circular orbit it needs only to maintain its velocity.*

d. *The speed of the planet in the circular orbit is the same at all points because its angular momentum constant, therefore because it is a circular orbit, its angular and tangential velocities are constant. Because the sector velocity at all point is equal, the speed at E > the speed at B > the speed at A for the planet in the elliptical orbit. The planet in the elliptical orbit has to slow down if it is going to be shifted to the circular orbit at E because the speed originally has is too much to remain in a circular orbit. The gravitational force of the Sun is not enough to pull it into the circular orbit at is initial speed at E.*

4. A space-communication company is planning to take a spy satellite to a spot 35,800 km above the Earth's surface and release into a geosynchronous orbit. (In a geosynchronous orbit the satellite will orbit at the same rate as the points on the surface of the Earth below it, so as to hover over the same spot.) Is this possible? If yes, how fast must the satellite be moving when it is released?
<u>Section 12-3</u>

Although this question asks for a direct application of Kepler's third law, it affords an opportunity to re-derive the law for a circular orbit on a concrete example.

Points to emphasize: Given the size of the orbit, the period is determined by physical laws. A geosynchronous orbit must be circular. (Note that the definition requires more than a 24-hour period.

The satellite has to remain overhead. This requires a circular orbit in the equatorial plane. This is not obvious to students until it is pointed out.) To three-digit accuracy, the radius of a geosynchronous orbit is 4.22×10^7 m. Thus, the above altitude is correct to the same three-digit accuracy.

PUZZLES

"The Orbit Paradox"

Consider a satellite in a circular orbit about the Earth. If NASA wants to move the satellite into a higher (circular) orbit, they have to increase the satellite speed and yet when the satellite is into the new (higher) orbit, its speed is actually less than it was in the old (lower) orbit. Is this correct? If the answer is yes, can you explain why the satellite slows down? Please answer this in words, not equations, briefly explaining how you obtained your answer.
<u>Section 12-5</u>

The total energy of the satellite in a higher orbit is more than its energy in the lower orbit. To increase the energy, the satellite's kinetic energy is increased, the centripetal force is too weak and the satellite moves away from the Earth. The satellite is now in an elliptical orbit. At the apogee of this orbit the satellite is moving quite slowly, having expended the energy boost to pull itself away from the Earth's gravitational pull. Without further correction, the satellite will continue on its elliptical path and return back to its original elevation. To place it into a circular orbit at the new, greater elevation, the satellite energy has to be increased again to give it sufficient speed to stay in the new circular orbit.

Here are some sample (unedited) student responses that provide useful starting points for discussion:

a. *Yes, I do believe that it does slow down. As you get further away from the Earth, the gravitational attraction doesn't have as much of an effect. There is a greater radius between the two objects. There are several equations that you could use to prove this. One that comes to mind immediately is the equation for velocity in a circular orbit. But I do wonder, is the satellite really slowing down, or is it just taking longer to complete the period?*

b. *Yes, this is true because when the satellite is close to the Earth the Earth pulls much harder on the satellite creating more speed so when the satellite is farther away from the Earth the satellite has less of a pull on it from the Earth.*

c. *This is correct. NASA boosts the speed of the rocket to fling it out into a higher orbit, but this orbit won't be circular unless NASA also slows it when it gets out to the desired orbital radius. The orbit must be slowed because the only force on the satellite is gravity which must equal the centripetal force that is keeping the craft in its circular orbit. As the orbit radius increases gravity gets weaker, so the satellite must compensate by being slowed. I think NASA actually slows it down. They fire the retrorockets or something.*

d. *Correct. Since the satellite is in a circular orbit, its speed is constant at a given orbital radius, and the only force acting on it is the gravitational force which provides a radial acceleration towards the object it is orbiting. The equation given for the speed of an object in a circular orbit is v = (Gm/r)^1/2/ Thus, an increase in the radius of the orbit results in a decrease in the velocity of the satellite. This is expected since the period of revolution increases with an increase in the radius. The speed is less in the new orbit because the increased radius results in a lower radial acceleration experienced by the satellite. In order to maintain a constant velocity given by the equation, this decreased radial acceleration causes the orbital velocity to slow down such that the radial acceleration can have a larger effect on accelerating the satellite toward the Earth.*

e. *Yes. The satellite slows down because if it were to maintain its original speed it would just continue into space, maybe in an elliptical orbit. To keep the object in a spherical orbit, you must slow it down to the appropriate speed, once it reaches the altitude that you want the orbit to be in. The earth's gravity has a smaller effect on the satellite at higher altitudes so a smaller velocity is needed to hold them there in their orbits. The closer the orbit the faster the velocity must be to keep the object from falling to earth, they just want the object to fall around the earth in a sphere using the earth's gravity to their advantage.*

Chapter 13: Oscillations about Equilibrium

WARM-UPS

1. Consider a mass on a spring, oscillating under the influence of a non-conservative retarding force, such as air drag. How would the retarding force affect the PERIOD of the oscillations? In a sentence or two, justify your answer.
Section 13-4

A retarding force would act against the restoring force of the spring. Throughout the motion the acceleration would be diminished (and no longer proportional to displacement). This means that under damping, the period will increase and the motion will no longer be simple harmonic.

Here are some sample (unedited) student responses that provide useful starting points for discussion:

a. *I don't think that the retarding force has any affect on the period.*

b. *The retarding force would increase the length of a period, but the amount of oscillations would stay constant.*

c. *The retarding force would decrease the period. It would cause the spring to work harder so that the distance traveled would be less each pass, making the Period decrease.*

d. *The force would have no effect on the period because the average acceleration of the oscillating object would be slowed, however the distance over which it oscillates (the amplitude) would be reduced.*

e. *The retarding force reduces the mechanical energy in the system. It will cause each cycle to take longer (i.e. the period will increase).*

2. When you push a child on a swing, your action is most effective when your pushes are timed to coincide with the natural frequency of the motion. You are swinging a 30-kg child on a swing suspended from 5-meter cables. Estimate the optimum time interval between your pushes. Repeat the estimate for a 15-kg child.
Section 13-6

If we approximate the system with a simple pendulum, the desired time interval is the period of a simple pendulum of length 5 meters, $T = 1.6$ sec. In this approximation the answer does not depend on the mass of the child.

If we approximate the system with a physical pendulum, the periods for the two children would be different. (Their centers of mass are different.) The correct estimate in this case would require the application of the parallel axes theorem, which is beyond the scope of this book.

Nevertheless, discuss the difference between the simple pendulum model and the more elaborate physical pendulum models. What happens if the child stands up in the swing?

3. Suppose you are told that the period of a simple pendulum depends only on the length of the pendulum L and the acceleration due to gravity g. Use dimensional analysis to show that the square of the period is proportional to the ratio of L/g.
Section 13-6

The unit for the period can be any of the units of time (second, minute, hour...).
Therefore the dimension of the period is T, (T for time). The other two basic dimensions in the SI system are M (for mass) and L (for length).
The length of the pendulum has the dimension of length (L).
The acceleration has the dimension (L/T^2).
The simplest way to combine the two quantities in such a way that a dimension of T results is by dividing L by L/T^2, and taking the square root of the result.
Thus the period must be proportional to the square root of the length divided by the acceleration g.

4. You want to construct a poor-man's amusement ride by mounting a seat on a large spring. Estimate the spring constant that would give you a ride with a period of 10 seconds.
Section 13-4

A plug-and-chug estimate.
k must be about 250 N/m for a 100-kg mass.
A chance to give students a feel for the magnitudes of these quantities.
What kind of amplitudes would we be talking about? Relate this to shock absorbers.

PUZZLES

"The Bungy Bucket"

Little Danny stands next to a large amusement park bungy bucket. The empty bucket is hanging on two large, identical springs. It oscillates with an amplitude of 2 meters and a period of 2 seconds. It is shown at the equilibrium position in the picture, 2.5 meters above the floor. Danny waits until the bucket comes to a momentary stop, ready to start its journey upward. He quickly takes a seat in the bucket. Danny's mass is equal to the mass of the empty bucket. What happens next? Does the bucket take Danny for a ride? (You can treat the two springs as a single spring with the spring constant k equal to twice the k of each single spring. In this estimate ignore the mass of the springs.)
Section 13-2

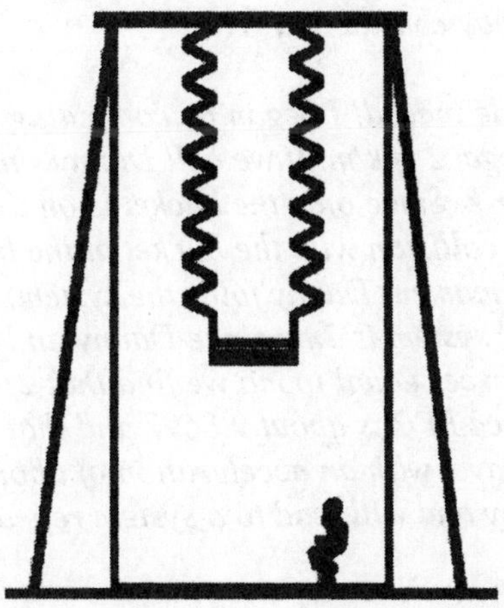

At 2 meters extension the restoring force of the spring is almost equal to the combined weight of Danny and the bucket. Thus the motion will stop or proceed with a tiny acceleration. (Note: Please see the algebra in the student responses below.)

Here are some sample (unedited) student responses that provide useful starting points for discussion:

a. *No, the bucket does not take anyone for a ride if their weight doubles the weight of the bucket. The reason for this is that the force of the springs is equal to the force pulling downward, and since there is no velocity the spring will remain in equilibrium at the ground.*

b. Given: A = 2m, T = 2s
The mass of the empty bucket = Danny's mass = m. The spring const. of the two springs combined = k
omega = sqrt(k / m) --> (2PI / T)^2 = k / m --> k = 4 * m * (PI^2) / T^2
After plugging in the value for T, we get: k = (PI^2) * m
At the lowest position, the force of the spring on the bucket is:
F = k * A = [(PI^2) * A] * m = 19.74 * m (upward direction)
After Danny takes a seat in the bucket, the combined weight of the bucket and Danny is:
w = 2 * m * g = 19.6 * m (downward direction)
The total force acting on Danny and the bucket right after he takes a seat is:
F(net) = F - w = 0.14 * m, which will finally result in a very small amplitude. Answer: Theoretically, the bucket will take Danny for a ride, though since the force pulling it up is minimal, that ride will be pretty boring.

c. The bucket does not take Danny for a ride. I analyzed a couple of equations and when Danny gets on the bucket it will sit on the ground. F=-kx where the force is the m*g of the bucket. Since the boy is the same weight as the bucket the force is doubled. If the force is doubled and k is a constant then x has to double. x=Acos(3.14t).The cos 3.14t is equal to +-1 at the endpoints. Since x doubles then the amplitude should double. If the bucket is at equilibrium at 2.5 meters above the ground and the amplitude of the bucket is 2 then its height above the ground is .5 meters. If the amplitude doubles then he doesn't go for a ride.

d. The bucket doesn't really take Danny for a ride. He and the bucket do oscillate, but with an amplitude of only 1.4 cm. To obtain this answer I first got the angular velocity from the given period. Then I used this angular velocity to obtain a spring constant of 9.87 kg/s^2 * M (M is the mass of the bucket.). Then, using a free body diagram of the bucket with Danny on it at the bucket's new equilibrium point (the unknown quantity), it is shown that kx = 2Mg (x being the change in the distance from the old to the new equilibrium point). Solving for x gives x = 1.986 m from the original equilibrium point, which is 1.4 cm from the point where Danny gets on. So the total that Danny will travel is 2.8 cm, hardly what a little boy considers a "ride."

e. *Little Danny will have a very slow ride indeed! I began by computing omega to be 2Pi(f). That is pi in this problem. Then we know that omega^2 = k/m. If we call Danny's mass and the mass of the bucket m, then we can solve this equation for k. Since only the bucket is on the spring when it descends, k=pi^2(m). If Danny has an inelastic collision with the bucket at the bottom of the oscillation as is described in the problem, then at the moment Danny joins the system, the upward force is kx or 2pi^2(m). The downward force on the system is 2mg since Danny and the bucket are both on the end of the spring. Setting the sum of the forces equal to ma we find that 2pi^2(m) - 2mg = 2ma. The 2 and the m cancel leaving pi^2 - g = a. Since Pi^2 is about 9.8697 and that is very close to g, Danny and the bucket only begin their upward travel with an acceleration of about .0696 m/sec^2. This new acceleration for the system is very tiny and will lead to a system repeating very tiny oscillations.*

Chapter 14: Waves and Sound

WARM-UPS

1. A certain transverse traveling wave on a string can be represented by

$$y(x,t) = A \sin (kx + \omega t)$$

where y is the displacement of the string, x is position of a point along the string and t is the time. What are the units of the coefficients of the x-term and the t-term in the above expression? What do we call those coefficients?
Section 14-3

In our experience, students have a hard time verbalizing these relationships. Thinking about the units gives them a start. Revisit the notion of phase. For a fixed x, kx is a phase factor in the y vs. t relation, for a fixed t, ωt is a phase factor in the y vs. x relation (note the second student's response below which is worth elaborating on. We would give the author of such a response a chance to elaborate).

Here are some sample (unedited) student responses that provide useful starting points for discussion:

a. *The coefficient of the x-term is called the wave number and the units are rad/m. The coefficient of the t-term is called the periodic wave, and the units are rad/s.*

b. *Those coefficients are called the 'phase'. They specify which part of the sinusoidal cycle is occurring at a particular point and time.*

2. Suppose you suspend a 3-m nylon rope from a hook in the ceiling and tie a 10-kg ball to the end of the rope. The mass of the rope is 0.5 kg. Now you pluck the rope. That sends a wave up and down the rope. Estimate the speed of propagation of that wave.
Section 14-2

Another formula calculation as a lead-in in the discussion of wave propagation and the role of the wave medium. The discussion can be as simple as a model with discrete masses coupled with springs or as elaborate as the derivation of the one-dimensional wave equation.
v for the system described in the question is about 24 m/s.

3. A certain transverse traveling wave on a string can be represented by

$$y(x,t) = A \sin (kx + \omega t)$$

where y is the displacement of the string, x is position of a point along the string and t is the time.
A wave can also be represented by a similar equation where the two terms in the argument of the sine function have opposite signs:

$$y(x,t) = A \sin (kx - \omega t) \text{ or}$$
$$y(x,t) = A \sin (-kx + \omega t)$$

Is this wave different from the one represented by the original expression? If the answer is yes, what is the difference?
Section 14-3

The algebraic signs in the arguments of the sine function determine the direction of propagation of the wave. If both terms have the same sign, the wave propagates in the negative direction, if the arguments have opposite signs, the wave propagates in the positive direction. Students often memorize this fact, but they are usually not able to explain. A reasonable amount of class time devoted to this issue is time well spent, particularly if students gave it serious thought. Relating a formula like this to the phenomenon it represents is a skill that can only be developed in a subject where mathematics is used, not in a math class.

4. How much faster does sound travel in steel than in water? What properties of the two materials are responsible for this difference?
Section 14-4

About four times faster. Another look at the effect of the properties of the medium on wave propagation. Density and elasticity have an effect on the speed of mechanical waves. While steel is denser than water, its elasticity more than compensates for the higher density.

PUZZLES

"Little Sally"

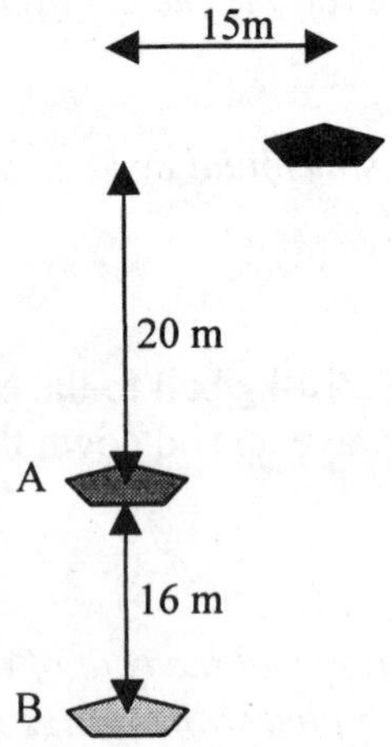

Little Sally is floating peacefully in her small red plastic boat, at the position marked "A" in the figure above. The lake is calm and the day is sunny. A motorboat speeds by, about 30 m off the shore, and the waves reach the shore about half a minute later. Suddenly, little Sally's older brother and sister decide to give Sally some fun. Each sibling is in a rowboat firmly tied to moorings out in the lake, as shown in the figure. They each jump up and down in their respective rowboats, bobbing up and down in synch together, about 15 times every half a minute. Soon large waves are emanating from each of the bobbing boats, and the older brother and sister are giggling fiendishly. Sally's father, seeing the situation and having taken physics, does some quick thinking and moves Sally to the position marked "B" in the figure. Sally likes to float in calm water. Does Sally's father do Sally a favor by moving her?
Section 14-7

At point B the waves from the two boats interfere constructively and Sally is not happy.
The speed of the waves is 1 m/s. (The wave from the speedboat 30 meters from the shore reaches land in 30 seconds.) The siblings' frequency is ½ Hz. (15 vibrations in 30 seconds.) Sally's boat is 26 meters from one of the siblings and 30 meters from the other. The waves reach Sally's boat in phase.

Chapter 15: Fluids

WARM-UPS

1. The atmospheric pressure at sea level is 1.013 x 10^5 N/m². The unit of pressure N/m² is also called a pascal (Pa). The surface area of an average adult body is about 2 m². Calculate this value in pounds using the fact that the weight of a 1-kg mass is 2.2 pounds. How much crushing force does the atmosphere exert on people? Why don't we get crushed?
Section 15-2

About 45,000 lbs. This gives students a feel for the magnitudes of unfamiliar units. The question opens the door to a discussion of deep-sea diving, pressurized airplane cabins, iron lungs, drinking through a soda straw, etc.

Here are some sample (unedited) student responses that provide useful starting points for discussion:

a. *Since we have a mass, we exert an outward force on the air pressure. Thus we counteract the atmospheric pressure so we don't get crushed.*

b. *About 45,482 lbs. There is an upward force under our feet equal to the downward force of atmospheric pressure equaling zero net forces if we neglect our weight.*

c. *We don't get crushed because our body is designed to push back as much as the atmosphere pushes in. Many of the things in our body really aren't that compressible, like water, and bones. Maybe that is why water is in everything in the body, so stuff won't get squished, or perhaps not.*

d. *I think it's because our atmosphere can be considered a 'sea of air'. This sea of air can be thought of as fluid in nature. There does exist a pressure that completely surrounds us, but it's uniform throughout, so we don't crushed.*

e. *The crushing force is simply the area times the pressure. We don't get crushed because this force actually helps hold our bodies together.*

f. *202,600N. We don't get crushed because humans are composed of 95% water which is virtually incompressible.*

2. The atmospheric pressure decreases with altitude exponentially according to the relation $p = p_o e^{-(0.00012)h}$ where h is the altitude above sea level in meters and p_o= 1.013 x 10^5 N/m². The average atmospheric pressure on a beach in Florida is close to p_o. Estimate the average atmospheric pressure in Denver. (The elevation at Denver is at about 1500 m.)
Section 15-2

About 84% of the sea level value, or 0.85 x 10^5 N/m².

3. Water is pumped up a pipeline as shown in the below figure. The water pours out at the top. The pump is running at constant speed. Compare the water speed at the three points A, B, and C in the pipeline. STATE YOUR REASONS FOR ALL THE ANSWERS.

Section 15-7

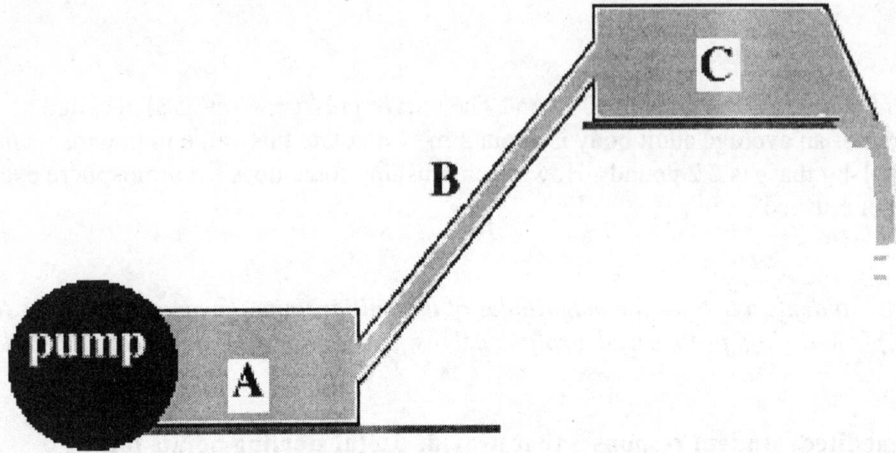

Because water is virtually incompressible, the volume flow must be the same at all points in the pipeline. Water flow must be faster at B because of smaller pipe cross-section. The flow speeds are about the same at points A and C. This is another non-intuitive topic. Once students accept the incompressibility argument, the other variables have to be dealt with: gravity, work by the pump, kinetic-energy change. Explicitly addressing all these issues will make the Bernoulli equation more palatable.

Here are some sample (unedited) student responses that provide useful starting points for discussion:

a. *a is the second fastest due to the pressure exerted by the pump*
c is the fastest due to the pressure of the pump and the force of gravity.
b is the fastest since it is working against gravity

b. *The water speed at point B is the fastest since its area of a cross section is the smallest. This is assuming that the water is incompressible. The water speed at A and C would be the same since their area looks to be the same.*

c. *The speed of the water at point B is the greatest and the speed of the water at point C is greater than (or maybe equal to) A. This is because the water starts from a stopped position at point A and goes gradually up hill to the maximum point where the water slows again to a stop and turns around.*

4. Estimate the force on a 1.5-m² windshield as a 60 mph (30 m/s) wind blows near its surface (parallel to the surface). The air in the car is stationary, and the car windows are shut tight.
Section 15-8

This is a highly non-intuitive application of the Bernoulli equation. The argument is similar to the one used in calculating the lift force of an airplane wing with air flowing at different speeds above and below the wing. The fairly standard (erroneous) argument compares a point below the wing (or inside the car) to a point above the wing (or outside the car) as if these were along the same flow lines. The correct argument compares the two points to a third point along common flow lines. From the two applications of the Bernoulli's equation, a pressure difference can be calculated, effectively giving $\Delta p = 1/2\rho(v_1^2 - v_2^2)$. *Using this argument, we get 870 N (196 lbs) for the force on the windshield.*

PUZZLES

"Sink Or Swim"

A rectangular block of wood floats submerged, 70% in water, 30% in oil, as shown in the picture.
What happens if you add some more oil?
What happens if you add some more water?
What happens if you siphon off the oil?
Do the percentages change? Which way?
<u>Section 15-5</u>

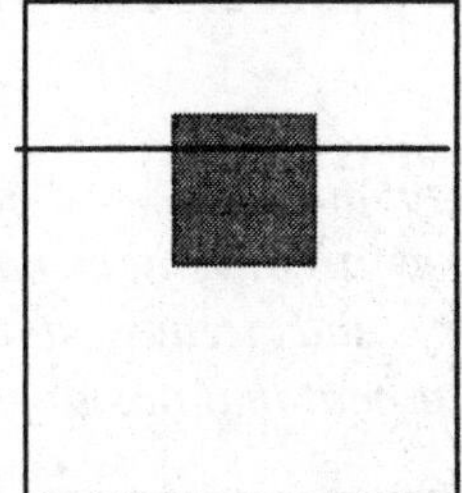

The buoyancy force F_B on the wooden block equals $F_w + F_o$ where F_w is the buoyancy force exerted by the water and F_o is the buoyancy force exerted by the oil.

$F_w = \rho_w V_w$ where ρ_w is the density of water, and V_w is the volume of water displaced by the wood.
$F_o = \rho_o V_o$ where ρ_o is the density of oil, and V_o is the volume of oil displaced by the wood.
The ratio ρ_w / ρ_o is 7/3.

If oil or water is added to the container, the equilibrium will not be disturbed, and the block will still float 70% in water 30% in oil.

If the oil is siphoned off, water will have to provide the entire buoyancy force and the block will sink lower into the water. It will now float 83% submerged (70% +3/7 of 30%).

Chapter 16: Temperature and Heat

WARM-UPS

1. A piece of wood at 130 degrees C can be picked up comfortably, but a piece of aluminum at the same temperature will give a painful burn. Why is this?
Section 16-6

This is a fairly easy question, and most students have a pretty good idea of the correct answer. Nevertheless, it makes a convenient point of departure in discussing the ideas of thermal conductivity, thermal equilibrium, and steady state processes. Further, a minority of students do have some misconceptions about heat that this question brings out. One might discuss the following set of responses:

a. *Aluminum is a good heat conductor. Wood is more of an insulator.... Heat can flow more easily through the metal to your hand, than heat travelling through wood.*

b. *It is because the piece of aluminum transfers heat more efficiently than the piece of wood.*

c. *Because their specific heat capacity is different. A piece of aluminum can contain more heat than a piece of wood at the same temperature.*

Here, the first response is the best. It includes not only the difference in conductivity but an understanding of why this is important. (Heat can flow more easily through the metal to your hand) The second is pretty good, but suggest that the student may believe that it is the hand-object interface rather than the bulk properties of the object that are important here. The third choice illustrates a more severe problem, which needs to be gently corrected. Although this misconception is not too common, a significant minority of students do exhibit it (perhaps 10%).

2. Estimate the amount of heat necessary to raise your body temperature by 1 degree Fahrenheit.
Section 16-5

This is a good example of an estimation question that requires "personal" data. It brings home the "real world" aspects of the subject when the object acted on is the student herself. Most students do a fairly good job with this question, varying primarily in the degree of sophistication used in estimating the specific heat of a human body. Here are two examples:

a. *1 degree Fahrenheit = .556 K (approx.) assume mass = 80 kg assume your body is 'mostly water' say 85% .85*80 = 68 kg 4191*68*.556= 158 kJ add 12 kg times my guess of 1500 for the specific heat of everything else that makes us not be a puddle. You get: 168 kJ (approx.).*

b. *Assuming my weight right now is 180 lbs (mostly water) = 81.6 kg. To determine how much heat is necessary to raise my body temperature by 1 degree Fahrenheit or .56 Kelvins. I used the formula Q=mc(delta T) Q=(81.6kg)(4190 j/kg K)(.56K)=1.9 x 10^5.*

3. A metal plate has a circular hole cut in it. If the temperature of the plate increases, will the diameter of the hole increase, decrease, or remain unchanged?
Section 16-3

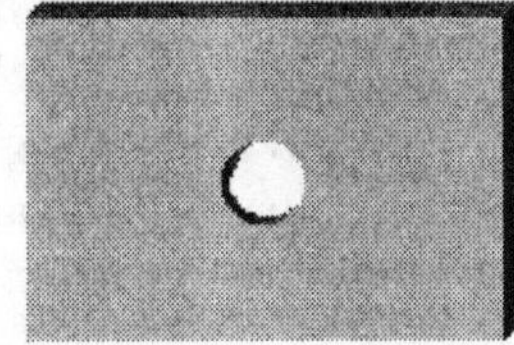

Many students expect that the diameter of the hole will decrease as the "surrounding metal" expands. In class, I often rephrase the question to have a square hole, then begin the usual derivation. At the end, I return to the round hole and ask the class to vote on the result.

4. In an attempt to open a new jar of peanut butter, you run very hot tap water over the (steel) lid. Estimate the change in the diameter of the lid.
Section 16-3

A nice opportunity to discuss the physics behind a common activity. Many students may have done this without realizing why it works. Also, many students are unsure about using a "linear" formula for an object that that is not "long and skinny." This example allows one to stress the use of the linear expansion formula for any linear dimension of any object.

PUZZLES

"Pot Physics"

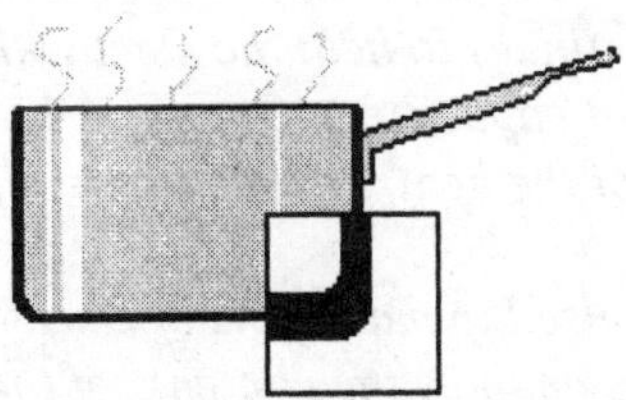

Advertisements for some expensive cookware have diagrams that look much like the figure above. The diagram is labeled to indicate that the inner and outer surfaces of the pot are made of stainless steel, and the inner layer is made of another metal such as aluminum or copper. Please explain why pots are made this way. Discuss why the different metals are chosen, and why a layered structure is better than using one metal and the same total thickness.
Sections 16-3 & 16-5 & 16-6

This puzzle helps get students thinking about heat transfer in an unfamiliar geometry. Pots are made this way because the stainless surface is strong, inert to most foods, and easy to clean; however, it is a poor thermal conductor. The inner layer provides of good lateral thermal conductivity so the inner surface has a uniform temperature profile that provides even cooking. Most students understand part of the value of stainless steel, but they have a wide variety of misconceptions about it as well. For instance, many students believe that the low thermal conductivity of stainless is important to protect food or chef from burns (see responses a–c). Others believe that the stainless layer provides "insulation," allowing the pan to retain heat (responses d–f). Few students understand the notion that the heat transfer must occur laterally, and focus entirely on the flow through the thickness of the pan. It is common for students to suggest that the inner layer makes the pan "heat up faster." That is, they imply that the low resistance in series reduces the total thermal resistance. A few students will take an entirely different tack and consider the differences in thermal expansion of the metals (responses g, h). In this set, response i is typical of good responses. I use a discussion of this puzzle to remind students that in the formula for conduction, the cross sectional area A and length L are determined by the direction of heat flow, not by their perception of which dimension is "long."

Here are some sample (unedited) student responses that provide useful starting points for discussion:

a. *Stainless steel is a less adequate conductor than Al. Since this is true stainless steel is placed on the outside of a pot to avoid burns to those who carelessly use the cookware. Al is placed on the inside of the pot because it can conduct heat much faster than if the whole pot was made of stainless steel. Finally stainless steel is just more durable than Al. This gives it a much more attractive and clean look than Al after a few months of being worn.*

b. *The pot is designed this way to more efficiently use the heat from the burner. If you use a normal aluminum pan food will not cook evenly and has a higher likelihood of burning. This is due to aluminum's high level of thermal conductivity. The heat passes through the pot very quickly and then passes very quickly into the food. Steel, on the other hand, passes heat more slowly than aluminum. This is beneficial in the fact that it won't pass on the heat of the burner into the food, and this lowers the likelihood of burning. However it takes much longer to heat up and takes longer to cook food. So by combining the two metals you get the benefits of both metals. Since the food is in contact with the steel it won't burn easily, and with the aluminum inside, the pot will heat up quicker and continually pass on the heat to the steel and on to the food.*

c. *The stainless steel gets very hot and distributes the heat evenly. Also the aluminum or copper can get so hot that it may melt, but continues to heat the steel which heats the food (since aluminum and copper have lower specific heat capacities). One thick layer of stainless steel would keep more of the heat at the bottom, or source of the heat, and wouldn't effectively heat the food.*

d. *The inner layer (aluminum) is used to make pots because aluminum is a good conductor of heat. Since it conducts heat well this enables the contents of the pot to get warmed up. The stainless steel enables the contents of the pot to remain warm. Different layers act as conductors and insulators.*

e. *One metal probably heats up quicker and the other metal is probably a conductor that retains the heat. As for the layers of metals it is probably better to have layers because they can 'insulate' better. Another way of keeping the food warm and at the right temperature as opposed to heating the food higher than normal and then cooling it faster than what is called for.*

f. *For a pot it is important to have heat to cook the food. The reason for having the different types of metal is because heat is held better this way so the pot stays hotter and at a better cooking temperature longer.*

g. *Different metals have different rates of thermal expansion. When objects, such as the pot in this example, are placed against one another and are heated, they tend to want to expand at different rates. Looking at the equation on the bottom of page 465, delta L = alpha times the original length times the change in temp., we can find some things. First, the change in temperature is virtually the same when the pot is heated. Secondly, for all practical purposes, they are the same in length. So, we discover that the change in length is dependent on proportionality constants for the two different materials. So, when heated, these two materials which are held together, start to expand at different rates. As they expand, the atoms of the copper which is expanding at a greater rate than steel (looking at the coefficients of linear expansion chart on page 466) are causing an increased stress on the steel. This increased stress on the steel causes the atoms in it to have a greater number of collisions, causing an increase in the heat of it. This is virtually drawing in extra heat, using less of the outside source (the burner of the stove).*

h. *Different metals are chosen as they have different coefficients of thermal expansion. The inner layer is made of copper/aluminum as they both have smaller coefficients of volume expansion than stainless steel. This prevents the inner material from expanding more than the outer material. A layered structure is better than using one thick layer of one of the metals, as this prevents thermal stresses from developing. If the pot was made entirely out of one metal, the outermost surface of the pot that has direct contact with the cooker would expand more than the aluminum beneath it causing thermal stresses to develop. By having the pot made out of several metals, the layers of the various metals can glide over each other as they expand.*

i. *The reason cookware is made of three layers is the inner and outer layers are for durability and easy clean up. These layers are usually made of stainless steel, which doesn't corrode and is easy to maintain and keep clean. The middle layer is usually made copper or aluminum and is put there to distribute the heat evenly throughout the entire pot. This is made possible due to the high rate of thermal conductivness of copper and aluminum. This kind of pot made of three layers instead of one is better because it is less likely to get hot spots and burn food. It will also last longer.*

Chapter 17: Phases and Phase Changes

WARM-UPS

1. The ideal gas law says the pressure of a gas on its container walls depends on the temperature of the gas. Please explain why this is so.
Section 17-2

The first example below shows that some students need occasional reminders that what we deal with is a model of nature. They confuse the equation (which describes nature) with the thing being described. The second response below also occurs occasionally. Many students have a sense that "as molecules move faster, they need more room," but they do not connect this to the concepts they have learned (momentum, in particular). The last two examples below illustrate the range of better responses. Most students understand that as T increases, the frequency of collisions with the walls increases (as in example c). However, few also understand that the average momentum transfer for each collision increases as well.

Here are some sample (unedited) student responses that provide useful starting points for discussion:

a. *As the temperature increases, the pressure increases as well, due to the simple fact that pressure and temperature are proportional to one another.*

b. *When you heat up a gas, the molecules gain energy and speed up and spread out so if enclosed in a container and unable to change its volume it will exert a greater force on the container walls trying to expand.*

c. *The higher a temperature is, the faster the molecules of the gas move. The faster they move, the more collisions they will have with each other and the walls of the container. The more the walls of the container are pushed on (or hit by the particles) the higher the pressure.*

d. *When the volume of the container is constant, and the temperature is increased, the gas molecules move more and faster, therefore they collide with each other and the walls of the container more often. They also hit harder, which creates the pressure.*

2. Estimate the number of gas molecules in 1 cubic meter of air in your classroom on an average day.
Section 17-1

This is also a good problem for giving students a feel for the numbers we are dealing with. Most students will get a correct answer, but many do so in a roundabout way. Most students are more comfortable with the ideal gas law using n and R rather than N and k, so they will use that method, then convert using Avogadro's number (as in example b, below). Some students will estimate the entire volume of the room, calculate a total number of molecules, then divide the volume out again. I will often take an example (such as c, below) to illustrate the dangers (too many steps gives more opportunity for mathematical errors).

Here are some sample (unedited) student responses that provide useful starting points for discussion:

a. *Using the equation pV = NkT, and estimating the temperature to be around 20 degrees Celsius: (1.013*10^5 Pa)*(1 m^3) = N*(1.38*10^-23 J/K)*(293 K) N is equal to 2.5*10^25 molecules.*

b. *Using the ideal gas equation...pV=nRT, or n=(pV)/(RT) n=(1E5Pa*1m^3)/(8.3*292K)=approx. 41 moles which is 2.5 x 10^25 molecules*

c. *Let's say that the room is 10X10X15ft. The volume in meters is then, 40.5m^3 so, 40,500L. The room T=25C, being 294K. And atmospheric pressure would be 1.013X10^5Pa. To find # of moles of air use n=PV/RT n= 1.47E11. 1mol has 6.02E23 molecules. Total molecules in the lecture hall is 8.87E34 per cubic meter there are 5.91E31 molecules.*

3. Name a correct unit for STRESS in the everyday pound, foot, second system. Name a correct unit for STRAIN in the everyday pound, foot, second system.
Section 17-3

Students are often confused by the stress vs. strain relation. I believe this is an example of a broad category of difficulties. Students are often confused by familiar relations when we make the transition from extensive to intensive parameters. In this case, the familiar Hooke's law to the unfamiliar stress strain relation. A similar difficulty occurs when Ohm's law is stated as E = J *rather than* V = IR. *This problem is intended to bring this connection out, and to encourage students to think about what each of these quantities means. Most do a fair job, although some manage to map SI units in English units without understanding the connection. Other students are uncomfortable with unitless quantities. The most common misconception is that these quantities have the same units.*

Here are some sample (unedited) student responses that provide useful starting points for discussion:

a. *Stress is F/A so its units are lb/ft^2. Strain is just (Lo-Lf)/Lo so it has no units.*

b. *Stress can be expressed as pounds per square inch and strain is dimensionless, so it has no units.*

c. *Stress is Lb per square inch. Strain is in percent Stress would be pound per square foot. The unit is psi.*

d. *The strain just describes the resulting deformation.*

e. *Pounds-per-square inch would be appropriate for both, stress and strain.*

f. *The units for stress and strain are both pounds per square inch*

4. Estimate the speed of a typical molecule of oxygen in your living room. How about a nitrogen molecule? A water molecule?
Section 17-2

This is a fairly easy problem, and most students can do it. I usually assign it in order to give students a better feel for the scale of velocities that we deal with, and in order to introduce other discussions (such as RMS vs. other averages, and the huge difference between molecular speed and "wind speed," which I bring up again in the context of electrical conduction). The only difficulty students face with this problem is recalling that some elements are naturally diatomic, as in the answer below:

The temp in my house is approx 70 F or 21.1 degrees C, 294 K. I find the mass of molecules, in kg from knowing their amu's off the periodic table. (Divide by Avogadro's, divide by 1000.) O: 2.66 x 10^-26 kg, N: 2.33 x 10^-26 kg, H2O: 2.99 x 10^-26 kg using V=[3kT/m]^1/2 gives V(O): 676 m/s, V(N):723m/s, V(H2O): 638 m/s

PUZZLES

"Vacuum Sucks!"

A vacuum chamber designed to be pumped down to a pressure of 10^{-4} Pa has stainless steel walls that are 3 mm thick. You must design a new chamber that can be pumped down to a pressure of 10^{-5} Pa. How thick must the walls of the new chamber be?
Sections 17-1 & 17-2 & 17-3

Many students have a notion of vacuum as a separate entity, rather than simply viewing it as low pressure. This is very similar to the idea of "cold" as independent from lack of heat. This puzzle challenges that notion directly. Many students will express the belief that the walls have to be "much thicker" but have no idea how to calculate it. Others will guess at the calculation. In contrast, some students will recognize that the forces on the walls are essentially the same, because the pressure differential is one atmosphere in both cases. I recommend beginning the classroom discussion by contrasting one well-written example of each type. Then, I would turn to the physics of the situation. What causes pressure? How many collisions/sec can be expected on a wall? How does that vary with pressure? To extend this puzzle, and to remind the students where their original ideas originated, it is worth talking about pressure vessels, in which heat walls must be used.

Chapter 18: The Laws of Thermodynamics

WARM-UPS

1. Explain how it is possible to compress a sample of gas to half its original volume without changing the pressure.
Section 18-3

Many students are confused by the jargon (compress is synonymous with decrease the volume) and this brings that out. More important, this problem offers a wonderful introduction to the whole business of thermal processes. Many students will base their answer to this question on the ideal gas law, without really understanding what needs to happen to effect an isobaric process. Others will think about the process but may come to incorrect conclusions about what must be done. Here are a few examples:

a. *If the volume is half as much, for pressure to remain the same, the temperature would have to be reduced by half as well.*

b. *By making a system that would allow you to use an isothermal process. That is, compress the gas very slowly and remove kinetic energy from the gas by cooling at the same rate that you add energy to the system. In other words, holding the temperature constant.*

c. *To hold pressure constant while decreasing volume, the temperature will have to be decreased. Both of these can occur simultaneously in such a way that we could reduce the volume without changing pressure.*

In this set of three, the first is correct, but contains too little detail. It is hard to know if the student could explain what must be done with the piston to effect the change at constant pressure. The second is more thoughtful, but incorrect. This is a common confusion—connecting particle energy with pressure. The third is an example of a good response. Contrasting these in class, and asking for a discussion of the strengths of each can get a good conversation going.

2. Estimate the change in temperature when one cubic meter of air is compressed to half its original volume adiabatically (start at room temperature and pressure).
Section 18-4

This question provides an opportunity to discuss the many practical systems that involve heating during an adiabatic compression, e.g., air conditioners and diesel engines. Many students will do this correctly, but many others are confused by adiabatic processes and will tacitly assume that the process is isobaric. Here are examples of each type:

a. *For this problem I will use 295 K as room temperature and 1.4 as the ratio of heat capacities. Using the equations in the text and solving for T2, I estimate the final temperature to be about 389 K or an increase of 94 K.*

b. *Starting at 300 kelvin and pressure at 1.01*10^5 pa and decreasing the volume of 1 m^3 in half would change the temperature to 150 kelvin.*

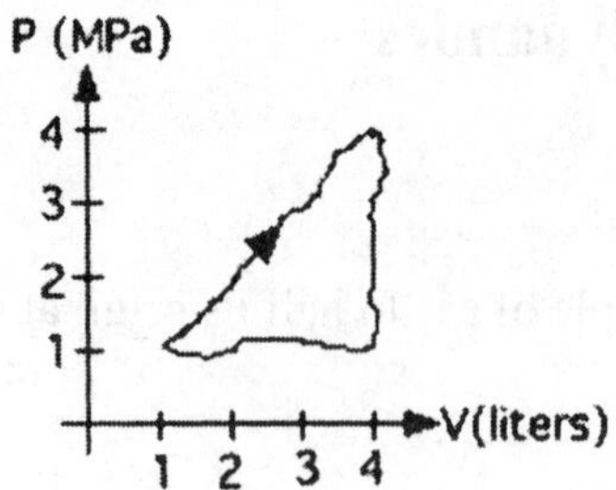

3. Estimate the total work by a system taken around the cycle shown in the above figure. Does the system have to be an ideal gas?
Section 18-3

This problem tests a student's conceptual understanding of work. There is no formula, and the process illustrated cannot be completely analyzed in terms of familiar processes (isobaric, isothermal, etc.) Furthermore, many students tend to get sloppy with their units when presented with graphical data. This problem also brings out students' misconceptions about what must be zero for a closed process. Many students believe that "everything" is zero for a closed process rather than just the total change in energy.

Here are some sample (unedited) student responses that provide useful starting points for discussion:

a. *Since the pressure and the volume are equal at the point of inception and the point of completion, no work at all was done. No, it doesn't have to be an ideal gas.*

b. *Total work = 0. Since there is no change in Volume from starting point to stopping point.*

c. *The total work would be the area enclosed by the path, which in this case is triangular. This triangular figure has two short legs which appear to have a value of '4', if the figure was a square then the area would be 4^2, since it is roughly half of the square then the area is approximately 8 units or 8J of work done.*

d. *The area of a triangle is A=(length*width)/2. So, from the picture, A=(3MPa*3L)/2=4.5MPa*L. Then 1L=1*10^-3m^3 So the total work done would equal 4500J.*

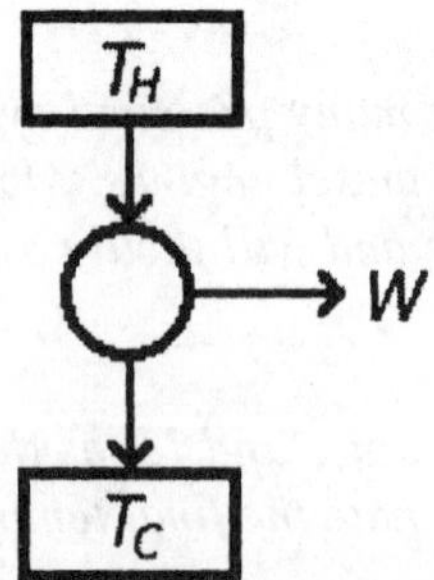

4. In your book, an illustration like the above is used to represent a heat engine. If this is to represent your car, what parts of your car correspond to the parts of the diagram?
Section 18-6

Many students have trouble with this question, but others do a good job. The figure itself is a great example of the kind of "schematic" diagram that physicists (and physics books) use constantly, but that many students find frustrating. This problem brings that frustration out in the open. The answers to this problem also present a good opportunity to discuss the ambiguity in the use of the word "system." Different students will give equally good answers to this question, while focusing on smaller or larger parts of the "system under discussion," the car. Here are some examples:

a. *The dark red at the top represents the combustion in the cylinder where the heat is produced. The circle in the middle represents some of the heat energy transferred into work done on the piston through expansion. The heat that is not transferred into work moves on to the radiator where it is cooled and the heat is lost to the surrounding environment.*

b. *The top part represents heat source, probably the combustion components of the engine. (gasoline?) The smaller, gray arrow represents work (this would be the energy that is transferred to the axle, or to the tires and to the road). The bottom represents the cold reservoir (this would probably be the radiator, using cooler air from surroundings).*

c. *The circle represents the engine. The top of the diagram is a reservoir (heat source) that supplies heat to the engine through the pipeline leading to the engine. The pipeline below the engine leads the cooled "heat" to the cold reservoir (the car's exhaust). This heat is discarded, "wasted." The branching line represents heat that the engine converts to mechanical work - energy with which the car will actually move and function.*

PUZZLES

"Process Problems"

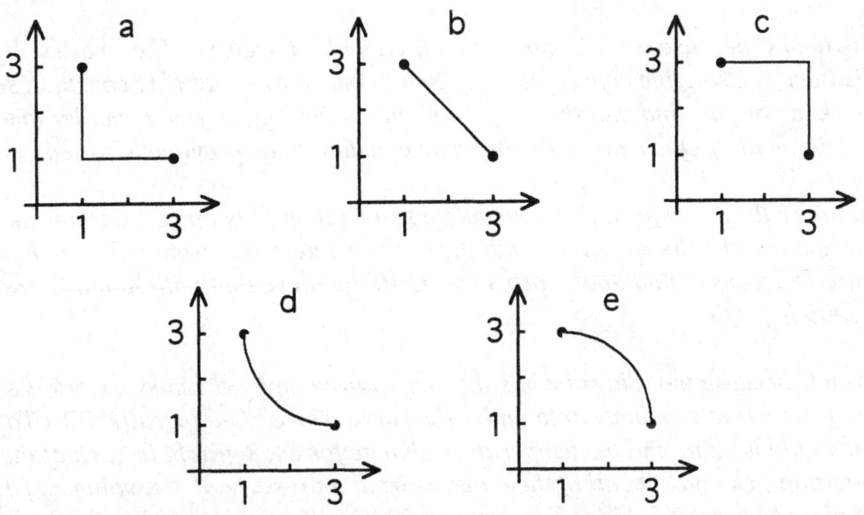

The above figure shows five different processes (labeled a-e) that each lead from an initial state (P = 3 atm., V = 1 m^3) to a final state (P = 1 atm., V = 3 m^3). Please answer each of the following questions. For which process is W the largest? The smallest? For which process is Q the largest? The smallest?
<u>Sections 18-2 & 18-3</u>

The most common mistake on this problem is for students to forget (or ignore) the minus sign that appears on the work in the first law. They understand that all of the processes must have the same change in internal energy, since they begin and end at common points. They also understand how to interpret the work from the graphical information. However, they then turn around and say that this means the process with the largest Q must correspond to the smallest W (responses a-c are of this sort). Another common error is for students to interpret process d as an adiabatic process and label it as the lowest Q, ignoring the possibility that Q may be negative, response d shows this error. Response e shows a less common flaw, but one that is worth mentioning in class. Many students latch on to the "mcΔT" category of formulas and have trouble letting go. They resist calculating Q in any other way. Responses f-h are all reasonable, and I would give them credit. However, I would also discuss a key difference between response f and the other two. That is, response f assumes the substance to be an ideal gas, which is not necessary. There are several useful extensions of this problem including ranking all of the choices, and estimating numerical values for Q and W in each case.

Here are some sample (unedited) student responses that provide useful starting points for discussion:

a. *Out of the processes that were given the 'a' pV graph is the smallest W, the area under the curve(line), and the largest Q. The 'c' pV graph is the smallest Q and the largest W, the area under the curve(line). My reasoning is delta U is independent of the path, but W and Q are dependent of the path. Delta U is therefore constant.*

b. *A has the smallest W (work) and the highest Q (heat gained or lost), conversely, C has the highest work and smallest Q. The reason is for the process between two points, change in internal energy (delta U) is path independent, so the change in internal energy for each of the PV diagrams is the same. Also then, the change in internal energy is equal to the heat gained or lost minus the work done by the system. So if work is the greatest for one such PV diagram, the heat gained must be the smallest so that the change in internal energy may be equal to the other paths*

c. *The largest W is item C because there is more of an area under the curve. The smallest W is item A. The largest Q is item A. Using the conservation of energy one knows that U is constant, so by plugging the work in you will find that the larger amount of work gives you a smaller amount of Q. and vise versa. The smallest Q is item C. By the same calculation as previously stated.*

d. *If W is the area under the curve for a P-V coordinate system then W is largest ondiagram C and is smallest on diagram A given the conditions that the curve is integrated from left to right. Q is smallest on figure D because in adiabatic processes Q=0. I believe that both A and C are largest and have equal magnitude for Q.*

e. *Work is largest in C because that diagram has the largest area under the curve, work is smallest in A because that diagram has the smallest area under the curve (since W=integral(PdV)) To find Q for each graph, you would have to find the temperature change for each graph. in each of the graphs, there is no temperature change. On all of them they start at a pressure of 3, a volume of 1, and end on a volume of 3 and a pressure of 1. Since Temperature is directly proportional to the product of pressure and volume, and the product of pressure and volume is exactly the same (3*1 =1*3) on both ends of the graph, we can see that there is no temperature change. Q is equal to m*C*delta(T). If there is no delta(T), meaning that delta(T) is equal to zero, then Q is also equal to zero. Therefore, Q is the same for all the above graphs.*

f. *Process a does the least work because the area under the graph is the least. Process c does the most work because the area under the graph is the largest. Because PV = nRT and since PV = 3 in the starting and ending states in all graphs, change in T = 0. Assuming this is an ideal gas, the internal energy depends only on temperature and since temperature doesn't change, neither does internal energy so Q = W. Process a transfers the least heat because it does the least work. Process c transfers the most heat because it does the most work.*

g. *Work is largest or most negative for C and smallest for A because C has the most area under it and A has the least. The largest Q would be the one with the most heat added to it. change in U=Q-W and since they all start out and end at the same amount of U, then the change in U must be equal if this is the case it means the one that does the most work requires the most heat. So, again C is largest and A smallest.*

h. *Since W is the area under the curve, we can see that W is the largest in "c" and the smallest in "a" Q is also the largest in "c" and the smallest in "a" The reason they are the same answers is because the change in internal energy for all of the graphs is the same, so Q-W is constant. Therefore, in order for them to equal out, when W is biggest, Q must also be biggest to add up to the same change in internal energy.*

Chapter 19: Electric Charges, Forces, and Fields

WARM-UPS

1. Can there be an electric field at a point where there is no charge? Can there be a charge at a place where there is no field? Please write a one or two sentence answer to each of these questions.
Section 19-4

Others have noted that many students believe the field cannot exist without something to "act on" as in the first sample response below. However, I have found that an even more common difficulty is students who cannot shed the notion of a"self field," as in responses b, c, and d. Response d is particularly interesting, in that the student would have given a correct response if the question had been phrased as a multiple choice, but the misconception is revealed in the more detailed response. Answer e, below, is one that I would show in class as an example of a particularly good response.

Here are some sample (unedited) student responses that provide useful starting points for discussion:

a. *If there is a charge at a point, there will always be an electric field. An electric field must have a point were there is a charge to exist.*

b. *Yes. There can be an electric field where there is no charge. Some charge at some other location can produce an electric field at this point. No. Any time there is a charge it creates an electric field in its vicinity.*

c. *There can be an electric field at a point where there is no charge. There must be a charge to create the field, but the charge does not have to be at that point, there is an electric field in the area around the charge. There cannot be a charge at a place where there is no electric field because the charge would create an electric field.*

d. *Yes; A point charge will create an electric field at that point and neighboring points as well. The neighboring points will have an electric field but not specifically a charge.yes; If a point charge is set in a place such that the electric fields of it and those causing forces against it cancel, there will be a charge at that point but no net electrical field.*

e. *Yes, there may be an electric field caused by a charge nearby that is emanating an electric field. Yes, it is possible for two electric fields of opposite magnitude to cancel each other out. Thereby creating a point without an electric field.*

2. Let's say you are holding two tennis balls (one in each hand), and let's say these balls each have a charge Q. Estimate the maximum value of Q so that the balls do not repel each other so hard that you can't hold on to them.
Section 19-3

At the beginning of the course on electricity and magnetism, it is crucial to start giving students a feel for the magnitudes of the charges, fields, etc. As the first two responses below indicate, many get through the mechanics course with little feel for a Newton. I would show the third response as an example of how to deal with this problem. It is also interesting to note how little students understand of our language. Since this is the first assignment with an "estimation" most students have seen, many do not understand how they are supposed to proceed as in response d. Similarly, response e was attached as a comment after the assignment.

Here are some sample (unedited) student responses that provide useful starting points for discussion:

a. *The answer to this question depends on how strong or weak the person in question is and how far the t-balls are placed apart. Since I personally don't know how many newtons of force I am able to exert, I'm just going to assume that I can hold on to a force of, say 10 newtons, and the balls are 2 meters apart. Then, Q would equals 3.33e-4 coulomb.*

b. *I'm not sure how many Newtons I can hang on to but I'm going to say about 100. If the distance between the balls is about .3 meters then 100 times .3^2 gives me 9. If I divide this by k I get 1E-9. Now I take the square root of this to find Q which is about 30nC.*

c. *The charge Q would have to be a value such that the force exerted by each ball on the other would be approximately 500N. This divided by gravity would be about 50kg (assuming I could hold 50kg in each hand). Thus the charge Q would be approximately 4 x 10^-4 c. (also assumed was arm length of 1.75m)*

d. *If it were as great as one C then it would be pretty hefty. Thus it would have to be a fraction of a Coulomb but larger than a micro Columb. Just a guestimation of course.*

e. *A little confusing when asked to give estimates. Should I assume things such as r being my out-stretched arm length/ I tried to work this out on paper just using the formula, but was not able. So I started assuming things about this problem to come up with an estimate.*

3. Is it possible for two electric field lines to cross? If so, under what conditions? Do electric field lines ever end? If so, under what conditions?
Section 19-5

For many students, electric (and magnetic) fields are the most abstract notion they have ever encountered. That we use multiple, inexact representations (field lines, vector maps, etc.) makes matters worse. Working with students' answers to this question creates an opportunity to discuss these concepts. Most students will have understood that the field lines do not cross, but may be unsure why. Many students do not understand the "creation and destruction" of field lines at charges, and this needs to be dealt with specifically.

4. Near its surface, the Earth has an electric field that points straight down and has a magnitude of about 150 N/C. Estimate the charge that you would have to place on a basketball so that the electric force on the ball would balance its weight.
Section 19-4

In general, the estimation problems are the most difficult for students, but also the most rewarding. This problem provides an opportunity to connect the new concepts of charge and field to ideas from mechanics and from their personal experience. Class discussion should stress that 150 N/C is a rather weak field, so the charge calculated is really quite large (compare to the two tennis ball problem, above). Incidentally, according to the Federation Internationale de Basketball (FIBA), the ball "shall be not less than 0.749 m

(74.9 cm) and not more than 0.780 m (78 cm) in circumference, and it shall weigh not less than 567 gr nor more than 650 gr."

PUZZLES

"Symmetry"

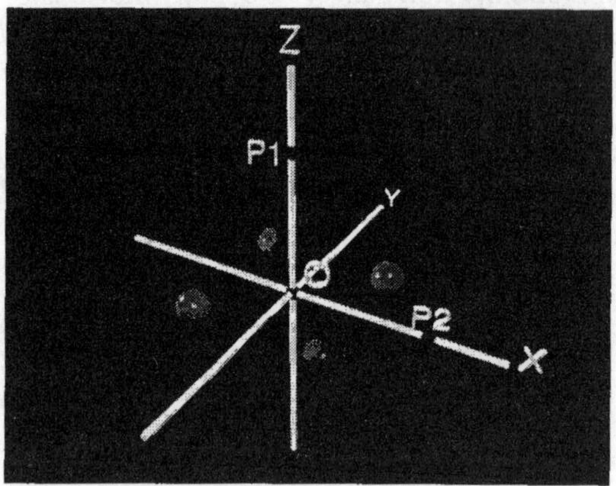

Consider the above figure. The interior balls are located at (-1,1,0) and (1,-1,0), and each carry a charge, Q. The outside balls are located at (1,1,0) and (-1,-1,0). They each carry a charge -Q. There are three points marked: O is the origin, P1 is (0,0,2) and P2 is (2,0,0). There are no charges at these points. I would like to rank the points O, P1, and P2 in order of decreasing electric field strength. The correct answer is:

a. P1>P2>O b. O>P1>P2
c. P1>O>P2 d. O>P2>P1
e. O>P2=P1 f. P2>P1=O
g. P2>P1>O h. P1=P2>O
i. The fields are all equal j. None of the above
<u>Sections 19-1 & 19-3 & 19-4 & 19-5</u>

This is a great question to test students' abilities to make symmetry arguments. It also can flush out a number of very persistent misconceptions about the electric field concept. When discussing this question in class, I first discuss the misconceptions (e.g., that there must be a test charge for a field to exist: responses a and b), then I move on to the symmetry aspects. Students make several common mistakes in these arguments. One common mistake is to say "the charges are symmetric so the field is zero everywhere." Response c is an example. Other students will completely ignore the vector nature of the field and argue that the field is highest at the origin due to proximity to the charges, as in responses d and e. A more subtle error is to see the symmetry at the origin, but to miss the symmetry at points on the z-axis, where no single pair of charges cancel each other, but the pairs cancel once all charges are included (response f). In this set, answers g and h are both good ones, which I would praise in class. It is easy to extend this problem by simply rearranging the charges. I have also concluded discussion of this problem by mentioning to students that this arrangement can be idealized into a quadrupole (as was previously done for a dipole) and that higher multipoles do exist and play an important role in advanced courses.

Here are some sample (unedited) student responses that provide useful starting points for discussion:

a. *I say the answer is i. Reasoning: By definition an electric field is the force per unit charge exerted on a test charge at any point. Since P1, P2, and O don't have a charge, they would all have equal electric fields.*

b. *The magnitude of the electric field is zero at all three points because if the charge is zero at that point then the magnitude and direction would be zero at that point.*

c. *They are all zero due to symmetry. The Origin is zero because the blue balls are attracted to the green ones. Since they are attracted equally on both sides, the blue ones will have a force toward the center resulting from the green ones. But the blue one on the other side resists the one on this side from moving closer. Therefore the particles will stay in their current position resulting in a zero value for the origin. The points P1 and P2 have a similar argument. They are symmetrically equidistant from charges that have the same absolute value. Therefore when doing the E calculation the field values from all directions add to zero.*

d. *I believe that the answer is b because the point O is in the fields of all 4 charges and closest to all 4. Next is the one on the P1. It is on the z-axis and that is the direction that the field points, and is equally close to all 4 charges. Finally, P2 is in the same plane but closer to two charges than the other two so I think that it is the least in electric field strength.*

e. *The answer is choice e, the distance from the origin to the charges is lesser than the distance from the charges to the points. The electric field is inversely proportional to the distance squared.*

f. *I would say answer g is correct for the following reasons. First, due to the symmetry of the arrangement, the field at point O is zero due to the cancellation of the forces at that point. Second, point P2 is the next weakest because of its position farther away from the charges. We know that E fields are affected by an inverse square law, meaning they decrease rapidly as the distance away from the charge increases. Thus leaving point P1 feeling the most influence from the E field emitted by the point charges.*

g. *The answer is f, P2>P1=0. To solve this I considered the field due to each charge that would act at the point in question. For point 1 and 0 if you break down the fields into components you can see that the net field is zero due to symmetry. For point 2, however, upon breaking the fields into components, the field components for the x direction cancel due to symmetry but the field components in the y direction do not and there is a net field in the + y direction.*

h. *At the origin, fields created by the positive and negative charges would cancel each other, field strength being equal and opposite (i.e. symmetrical). SO, the net field strength would = 0. At point P1 the green balls create an electric field in the -z direction. Due to symmetry the fields in the x & y directions cancel. Likewise the blue balls create an electric field in the +z direction with the x & y directional fields being canceled. Since at point P1 there is an electric field in the -z and +z direction of equal strength the total net electric field is 0. At Point P2, the charges around it are equal but there is not symmetry therefore an electric field will be present. In this case the electric field is in the +y direction. The answer is therefore F.*

Chapter 20: Electric Potential and Electric Potential Energy

WARM-UPS

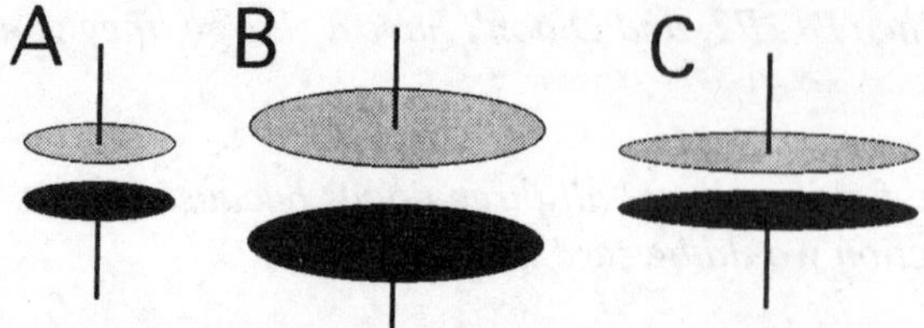

1. Suppose you are given three capacitors consisting of aluminum discs configured as in the figure above. You put equal amounts of positive charge on the top discs of each of the three sets and corresponding amounts of negative charge on the bottom discs. If you were to measure the electric fields between the discs, which set would give you the highest value? Which one the lowest? How about the voltages?
Section 20-5

Most students will have picked up that the electric field depends only on the charge density, and hence is larger for choice A, however, they have much greater difficulty with voltage. In particular, they have trouble balancing the effects of field and gap (response b, below, is a good example of a student wrestling with this issue). Responses like this are great for motivating the concept of capacitance as a way of bringing those two features together in a single parameter describing the effect of conductor geometry on voltage difference. The figure was drawn with the radius on plates B and C double that of A. However, students are often suspicious of reading too much into the diagrams. Response c shows a student who has found a way to deal with the ambiguity.

Here are some sample (unedited) student responses that provide useful starting points for discussion:

a. *The Capacitor A will have the highest Electric field, because it has a higher charge density. B and C will have the same field. However B will have a higher voltage than C.*

b. *The field of A would be the highest, because if we have the same amount of charge in a smaller area, the field is larger. The size of the E field has nothing to do with the distance that the plates are apart. Voltages - in comparing B and C the choice is easy, B wins because it has a larger separation between the plates. I can't really tell about A however, because I don't know if the decrease in area is sufficient to offset the decrease in distance relative to B. I think that they might even be equal.*

c. *I gave each plate a charge of 2. Discs A & C have the same d of 2. Disc B had a d of 4. Discs B & C had area of 4, disc A had area of 2. E=Q/(epsilon0*A). 'A' would have the highest value of E because it has the smallest area. B & C have the same E because their area is equal. Discs A & B have the same voltages using V=E*d. Disc C has half the voltage of discs A & B.*

2. Electronic flashes in cameras flash when a capacitor is discharged through a material that produces light energy when electric charge passes through it. This energy comes from a battery inside the flash unit. Estimate the capacitance of the capacitor used to store the energy. (Hint: Assume that the flash gives you the same amount of light as a 500-watt bulb that is turned on for a 60th of a second)
Section 20-5

This question is a bit "unfair," but it can be used to make a point. Most students will figure out the energy correctly from the hint, then correctly set the result equal to (1/2)CV2. What comes next is the unfair part.

These flashes actually charge the capacitor to several hundred volts (330 V is typical). Using 330 V, the result is about 150 F, however, if the student uses a typical "battery voltage" of 1.5 V the result is over 7 F! Many students will realize that this is an unreasonable answer, but some will not. This provides an opportunity to discuss estimation in general, as well as to discuss capacitance. A few representative examples follow. A few good students may look up the relevant voltage online.

Here are some sample (unedited) student responses that provide useful starting points for discussion:

a. *If we assume the battery in the camera is a 1 volt battery then the 8.3 J is proportional to 8.3 C which is also the capacitance since Q/V is capacitance.*

b. *I tried to solve this and I got an answer with a very large capacitance. But I know that Xenon strobes operate off of a very high voltage. (although the exact figures I don't know, it is in the 1000's of volts) So, the capacitance of the camera's capacitor is actually quite small, it just has a tremendous working voltage.*

c. *I think about 8.3 Farads - But in one of the examples in the book it says that this could not be right, that 1 Farad capacitors barely exist.*

3. A fresh "D cell" battery can provide about 5000 J of electrical energy before it must be discarded. Estimate the number of coulombs that must pass through it during its lifetime.
Section 20-2

This problem also helps students connect electrical concepts to more familiar ones, in this case, coulombs to joules at a "familiar" voltage. Further, many students never completely grasp the difference between energy and power when they first encounter it in mechanics. This confusion comes out again when the subject is less familiar, and many students will feel that a given battery must have a specific power. This notion can be dispelled during a discussion of this question.

4. The voltage across a capacitor is given by the formula V=Q/C, where Q is usually called "the charge on the capacitor." Where is this charge in a capacitor? Does the capacitor really have a net charge? If two capacitors are connected together is the charge on one the same as on the other?
Section 20-5

Many students are confused by the nomenclature we use. If a student does not prepare for class, he or she may take the phrase "charge on the capacitor" to be ambiguous, indicating the charge on one plate or the other. Asked to answer this question specifically, most get it right. There are a number of remaining misconceptions, though, as responses c and d (below) indicate. Students are unsure exactly where the charge resides, or why the capacitor must remain neutral. Students are less clear on the question o when two capacitors must have equal charge. Some understand (e.g., response a), but others (response c) do not.

Here are some sample (unedited) student responses that provide useful starting points for discussion:

a. *When people say that a capacitor has charge Q, it usually means that the plate at higher potential has charge Q, the plate at lower potential has charge -Q. The charge Q usually refers to the plate at higher potential which is the positive charge show on the bottom of the plate on the graph. The capacitor's real net charge is zero. Depends on how two capacitors are all connected; if they are all connected in series, they have the same charge; if they are all connected in parallel, they have different charges unless the capacitance is the same.*

b. *The capacitor as an entire circuit component, has a net zero charge. The charge Q refers to the magnitude of the positive charge on one plate, which is equal to the negative charge on the other plate. If 2 capacitors are connected, it depends on the type of connection if Q1=Q2.*

c. *A charge in a capacitor is uniform due to the conducting material that stores this charge on the inside. This causes the capacitor to have a net charge. In connecting two capacitors, the charge is not necessarily the same because they could have differing capacitances.*

d. *Since the capacitor is two conductors separated by an insulator, the charge is sandwiched inside the two layers. There is no net charge because of the nature of both the conductor and the insulator.*

5. You and a close friend stand facing each other. You are as close as you can get without actually touching. If a wire is attached to each of you, you can act as the two conductors in a capacitor. Estimate the capacitance of this "human capacitor."
Section 20-5

Many students have a big problem with this question. They know that the definition of capacitance is Q/V, but the notion is new to them, and they cannot see how to apply any of the other knowledge in the book. Some simply give up, as in response a, and other estimate a Q and a V (responses b, c). This gives a good opportunity to talk about the process of estimation. What is a "legitimate" assumption and what is not, and how to make the fewest assumptions possible. It is also a good time to stress that capacitance is independent of charge and voltage. That in calculating a capacitance the charge is always proportional to the voltage, so these cancel, leaving only the geometry of the conductors.

Here are some sample (unedited) student responses that provide useful starting points for discussion:

a. *The correct formula for capacitance is C=Q/V. So, we would have to know the charge (if any) of these people's bodies, and also the voltage(or potential difference) between them. Is this what you are asking for?*

b. *If the human body could hold .01 microcoulomb and the people were connected to a 5 volt source then the capacitance, assuming the people could be treated as parallel plates, would be 2.00 nanofarads.*

c. *C=Q/V If a charge of 10nC through a potential of 12V then the capacitance would be 8.3 x 10^-9 C/V.*

d. *In estimating the capacitance of this "human capacitor," I will make a few assumptions. First I will assume that my friend and I are acting as parallel plates. Second I will assume our "plate area," A, to be 1 square meter. I will also assume that our "gap" or distance, d, between us is 1 millimeter. Using the equation C=epsilion naught(A/d), the capacitance of this "human capacitor" is 8.85 x 10^-9 farads.*

6. Contour maps like the below show equally spaced lines of constant elevation at the Earth's surface. Please explain in what sense they are "equally spaced." Also, please explain in what sense these lines could be called gravitational equipotentials.
Section 20-1

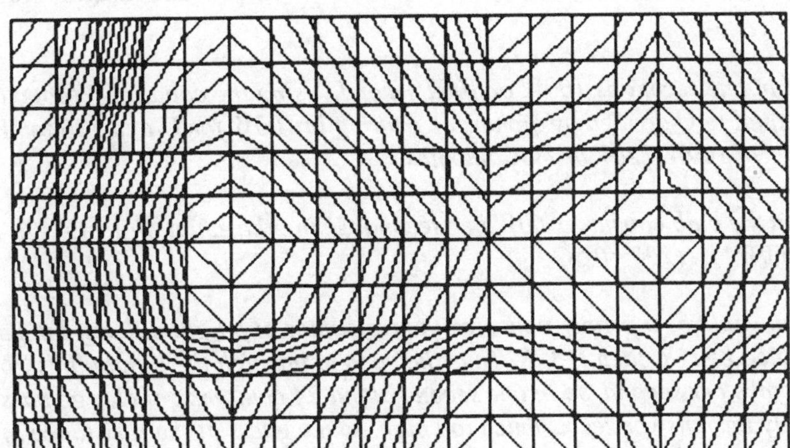

Here are some sample (unedited) student responses that provide useful starting points for discussion:

a. *Equally spaced in the sense that when the elevation is level the lines are spaced at a distance d. As the elevation changes that horizontal distance d becomes the hypotenuse of a right triangle (as seen in a cross section), with a height (or y component) that corresponds to the change in elevation and a horizontal (or x component) that make the lines look closer together.*

b. *The lines are spaced equally if elevation is constant in the sense that the rate of incline is constant and not accelerating (like walking up a quarter circle)*

c. *They are equally spaced in the sense that the points connected by the line are at the same, or equal, elevation. If a mass of M were moved along these lines, then the gravitational potential energy would not change since the elevation doesn't change. Therefore, the lines are also called gravitational equipotentials.*

d. *The lines are in a sense "equally spaced" because they represent three dimensions on a two dimensional surface. The dimension of elevation is represented through its change, that is to say a sharp change in elevation is represented by the lines being close together and a more gentle sloping is represented by a further spacing between the lines. If one were to roll a mass over a single contour line, the gravitational potential energy or mgy would not change because the elevation is constant. The greatest gravitational force would be realized where the lines are closest together and least where the lines are spaced further apart.*

7. Estimate the total charge that passes through a lightbulb in one second. (Hint: You can use "common" values for the voltage and the power.)
Section 20-2

This question also helps connect the abstract concepts in the book to the world of electrical devices that are familiar to the student. For this problem, the "Hint" is everything. I have found that with the hint above, most students get the problem fairly easily (see response a, b). However, if the hint is reduced, so that only the common voltage or power is mentioned, then most students are at a loss (responses c-f). A few do manage to get it, though (g).

Here are some sample (unedited) student responses that provide useful starting points for discussion:

a. *The energy used in one second is 100W or 100J/second. Using the equation V(potential energy per unit charge) is equal to U divided by q(charge). Putting in 100W for U and 120 for V and solving for q gives an answer of 100W/120V or .83Coulombs.*

b. *I will assume that my bulb consumes 60W per second. Using the equation P=Q*V/t, I can derive that Q=p*t/V -> Q=60W*1s/120V=0.5C*

c. *I'm not sure what would be a proper value for the resistance of a lightbulb but I'm going to guess about 5 ohms. Since V=I*R, then I=V/R. Thus the current through the lightbulb is about 24 A, which is given in Coulombs per second, so in one second about 24 coulombs of charge pass through the lightbulb. Actually this answer seems much too high, so my guess at the resistance of a lightbulb was probably off.*

d. *I don't know where to begin????????*

e. *Well I know that I need to integrate with respect to time the charge that passes thru the light bulb, but I am not sure how to go about it with the given information.*

f. *Well, frankly I've looked and read, and can't find anything to answer this question with just that amount of information.*

g. *The total charge that passes through a lightbulb with rating of 60 watts is .5 coulomb. I obtained this by using the formulas Power = V squared / R and power = I squared * R. I found R using the first formula and plug it into the second formula to get I. I is the rate of charges 'flow' (They don't really flow, they wiggle), and Ampere is coulomb per second. Therefore the charge that passes through the lightbulb is the current passing through the lightbulb and in this case, .5 ampere or .5 coulomb per second.*

PUZZLES

"Stuffed Capacitor"

Consider the following figures. On the left, is a plain parallel plate capacitor (area = A, gap = d). On the right is a system consisting of the same capacitor with a metal plate inserted into the gap. Assume the thickness of the plate is d/2. How will the new capacitance compare with the old?
<u>Sections 20-1 & 20-2 & 20-4 & 20-5</u>

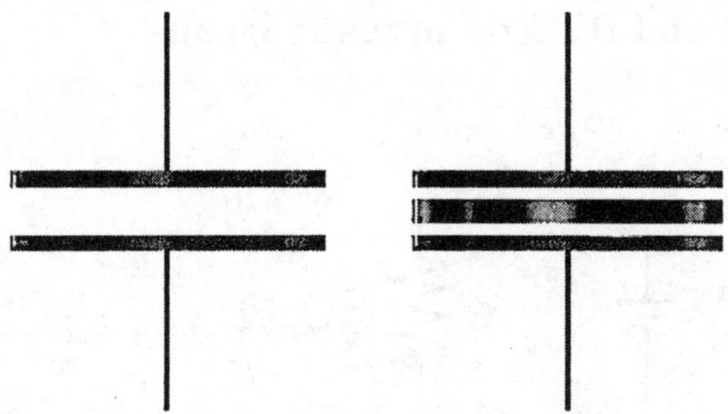

Most students will get this one correct, but it takes them some time. Eventually, they do figure out how to treat the "stuffed" capacitor as two in series. Those that do not often fail in a spectacular way (see the first two responses, below). Some students figure it out, but forget how to add capacitors in series, these come up with the result $C_2 = 8C_1$, as in response c. Responses d and e below are good ones. I particularly like to discuss this problem by noting the "two in series solution" as good, but then relating it directly, from first principles. It is also worth noting that it isn't necessary to assume that the plate be placed symmetrically in the gap. One very clever solution is to note that since the plate is uncharged there are no forces on it once it is in the gap, this means that one can move it up and down freely without changing anything. That implies that it can be placed in contact with one of the original plates, thus producing a single capacitor of area A and gap d/2. Extending this problem to a dielectric slab can be the beginning of a great discussion. This leads to an understanding of a conductor as an "infinitely polarizable" dielectric, and to a discussion of the forces on the slab as it is introduced.

Here are some sample (unedited) student responses that provide useful starting points for discussion:

a. *Most metals are conductors, so there is a conductor within a capacitor in the "stuffed" capacitor. Since this is a conductor, the electric charge would go through the voltage just as easily since the metal conducts this through. Therefore the capacitance which equals C*V would be the same. It all depends on the type of material inside the vacuum. It was an insulator, the capacitance would chance.*

b. *I think that the capacitance will be the same for both because the gap is still the same in both cases. I don't really know if the gap of the metal plate would have an effect on the outcome of the capacitance considering the fact that after it is all said and done we still have equally gapped parallel plates.*

c. *In effect I think that this would make another capacitor, so that now I have two capacitors in one, but their distances would be halved. (since the thickness of the plate is d/2) there is only d/2 of empty space left to divide between the other two effective capacitors. So that means that instead of D distance between the plates there is now only d/4 distance between 2 sets of plates. SO THE CAPACITANCE GOES UP 8 times.*

d. *Assume the metal plate is an ideal conductor. I now have two capacitors: each with area A and gap d/4. The original capacitance was epsilon0(A/d). I now have two capacitors in series, each with capacitance epsilon0(4A/d). Capacitors in series add reciprocally, so the total capacitance is epsilon0(2A/d), or twice what it was before.*

e. *I believe that the capacitance would have to double on the new capacitor. By putting a plate in between the plates, we have created two capacitors in series with area = A and distance equal to 1/4D. Adding these two capacitors in series, we get a new capacitance of 2 times the original.*

Chapter 21: Electric Current and Direct-Current Circuits

WARM-UPS

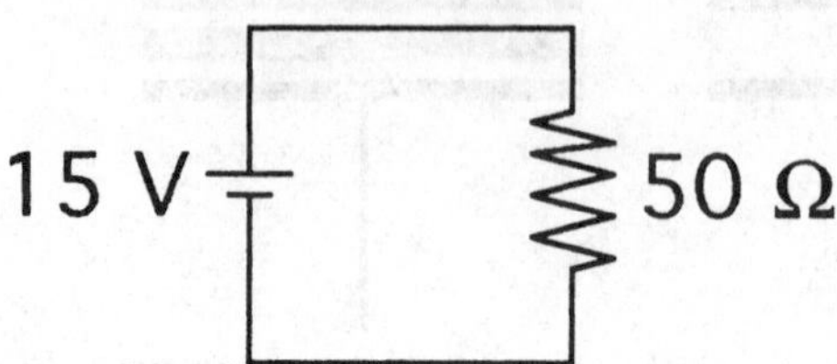

1. In the circuit shown, the battery provides a current I = 0.3 A to the resistor. How much current is returned by the resistor to the negative terminal of the battery? How much current flows inside the battery from the negative to the positive terminal?
Section 21-1

This question addresses two common misconceptions, that no charge flows inside the battery and that current declines inside a resistor. Many students believe that the battery is essentially a charged capacitor, with an excess of charge on one terminal and a deficit on the other, as the battery is used, this imbalance is relieved. These students are quite surprised that current flows from low potential to high potential inside the battery. Many students also believe that resistors "reduce" current internally. In a simple case such as that shown, they would argue that "none is left" for the return to the battery.

Here are some sample (unedited) student responses that provide useful starting points for discussion:

a. *The same current returns but I believe it also creates heat in the resistor. No, current doesn't pass through the battery. I believe the battery is similar to a insulator with plates.*

b. *The current is different because of the resistor on the right side. This is connected on here so that the battery does not get an overload. There is a voltage drop over the resistor. I'm not certain if the current necessarily goes through the battery. Batteries have anodes and cathodes, and charge isn't necessarily moved in the battery.*

c. *The same current does not return to the negative side of the battery because the resistor decreases the current. Current does pass through the battery.*

d. *The current does not pass through the battery because the plates are not physically connected.*

e. *The same current that goes in comes out. That same current does go through the battery, but how?*

f. *The current that returns to the negative side of the battery is less than the current that left the positive side because of the resistor.*

2. Estimate the resistance of a typical electrical cord in your home, say, one that attaches a lamp to the wall socket.
Section 21-2

Most students consider this a fairly easy question. It is a good start for establishing the connection between the textbook concepts and familiar devices. I often begin with this familiar result and move to the far more surprising calculation of the time it takes an electron to pass through the length of the cord at a typical drift velocity.

3. According to your textbook, the power dissipated by a resistor is given by $P = V^2/R$. However, Your textbook also says that the power dissipated by a resistor is given by $P = I^2R$. Is the power really proportional to R or is it really proportional to $1/R$? Can both be correct?
<u>Section 21-3</u>

This problem does a good job of getting students thinking, and provides a natural introduction to the topic of power. It also helps focus students' attention on the issue of which formulae are fundamental and which are derived for special circumstances. Most students are clear that neither "choice" presented in the question is strictly correct, but many are unsure what is correct. Here are some typical responses:

a. *Resistance is the reciprocal of the current, I, which is proportional to the power. So, the power is proportional to R.*

b. *At first glance the math says that both are true. But then when you think about it, power can't be both proportional and inversely proportional to the same thing. So you start thinking a little bit, and you realize that power really depends on what current you put through the resistor, not what its resistance is. And when you do that, you see that power is proportional to I squared.*

c. *I think both are true. When R=V/I it is the same as V=IR. When the later is substituted for V in the first equation it ends up being the same as the second equation.*

d. *Neither is really true, The electric power is proportional to the voltage or the current squared. In either of the formulas shown above the resistance is in the numerator or in the denominator. Thus showing that both are true.*

4. Estimate the resistance of a typical lightbulb. Hint: You can use "typical" values for the voltage and power.
<u>Section 21-3</u>

Students fare much better with this after a discussion of the previous "lightbulb question." However, a few will always try to do this by estimating the dimensions of the filament. Showing a couple of these in class is valuable because it opens the door to a discussion of how sensitive an estimate can be to a quantity that is difficult to estimate and which enters the calculation raised to a power (the radius of the filament). It also provides a natural opening to discuss the temperature dependence of resistivity.

Here are some sample (unedited) student responses that provide useful starting points for discussion:

a. *In a standard 120 volt circuit, a standard light bulb will have a resistance of 240 ohms. We get this number from the relationship the power is voltage times amps. A typical lightbulb will consume 60watts of power. Doing the math we see that it will also use .5Amps of current. Putting this into ohms law, we find that 120volts = .5amps * 240Ohms.*

b. *Assume the light bulb is 110 W, 120 V Rearrange P(power)=V(voltage)*I(current) to find current: I=P/V=.917 Then rearrange the equation that explains Ohm's law V=I*R and solve to get resistance: R=V/I=130.9*

c. *R=(row)*L/A. Row(for tungsten) = 5.25e-8. L=.02m A=.000001m^2 R=.0105 ohms which sounds wrong.*

d. *Assume L=0.01m (tungsten) and A=1e-6m A tungsten light bulb is a typical light bulb and has a resistivity of 5.25e-8. Resistance = rhoe*L/A=5.25e-8*0.01m/1e-6m = 0.000525 ohms or 525 micro ohms.*

PUZZLES

"Bulb Market"

Houses, boats, and cars have electrical systems based on 120 V, 24 V, and 12 V respectively. Let's say we have a 60-watt bulb from each, and enough adapters to plug any bulb into any system. Which combination of bulb and system would give the most light? Which would give the least?
<u>Sections 21-1 & 21-2 & 21-3</u>

Students get misconceptions about electricity from many sources, and the labeling on common electrical devices is one of the most common. All students know that lightbulbs are purchased "by the watt." More expensive bulbs provide more Watts, and halogen bulbs, which are extremely expensive, are proportionately bright. This question challenges this notion, and asks students to spend a few minutes determining what a power rating on a lightbulb really means. Many are surprised to find that high-power bulbs are really low-resistance bulbs. In the classroom, I recommend starting with a discussion of the ideal case, that is, the power "possibilities" assuming that these bulbs do not blow out. I would then move on to a discussion of more realistic labeling, which includes both a power and a voltage (or, in the case of resistors, a resistance and a power).

Chapter 22: Magnetism

WARM-UPS

1. The force on a charged particle in a magnetic field is very different from the force due to an electric field. Please list as many differences as you can. Don't forget to include differences in the direction as well as the magnitude.
Section 22-2

Once all the responses are in, most of the obvious differences have usually been listed (weaker since μ_0 is small, force dependent on v, force perpendicular to B). I will just show the list. However, there are a few subtleties that only one or two students note (sometimes none), and I like to highlight those (force zero if v is parallel or antiparallel to B). I also like to move from here to a discussion of how similar the forces really are (inverse dependence on r, proportionality to charge).

2. Estimate the force of the Earth's magnetic field on a 10cm segment of a typical wire in your home. (Hint: the magnitude of the Earth's magnetic field is about 5.5 x 10–5 T.)
Section 22-4

As the responses below indicate, most students are able to find an appropriate equation and use it. The difficulty lies in making an estimate of the current. Some do a reasonable job (e.g., response a), while others do not (response c). This latter category reinforces the need for estimation questions like the ones above that establish reasonable values for common devices. Response d in this set is a good response, and it brings out several important points. Many students have difficulty visualizing the cross product, and this should be brought out. I would also show this response because it makes the point that "Anyway, the force is not large."

Here are some sample (unedited) student responses that provide useful starting points for discussion:

a. *Let's assume that the current in the wire is flowing perpendicular to the magnetic field. I don't know what the current is in a common household wire, but let's set this value to be 2 amps. Using the formula for the magnetic force along a wire, the estimated force is 2 * 10^-5 Newtons. Assumptions: 12 gauge outlet type wiring with max carrying capacity of 20A, B and l are perpendicular so angle is 90 deg, and sin 90 deg = 1 given F = Il x B = 20A(0.1m) x 0.5 x 10^(-4)T = 0.002N*

b. *For my estimation I will use 60 amps for the current, I, and 5 x 10^-5 T for the earth's magnetic field, B. I will also assume theta to be 90 degrees. Using the formula F=IlBsin theta, the force would equal 6 x 10^-3 Newtons.*

c. *I am assuming my wire is straight and perpendicular to the Earth's magnetic field so F=IlB. The current is about 10A and the magnetic field is about 5 x 10^-5T so the force is about 5E-5N. I am not sure about the perpendicular part though. I am thinking it might be parallel in which case the force is zero. I am having trouble with perpendicular in a three dimensional world! Anyway, the force is not large.*

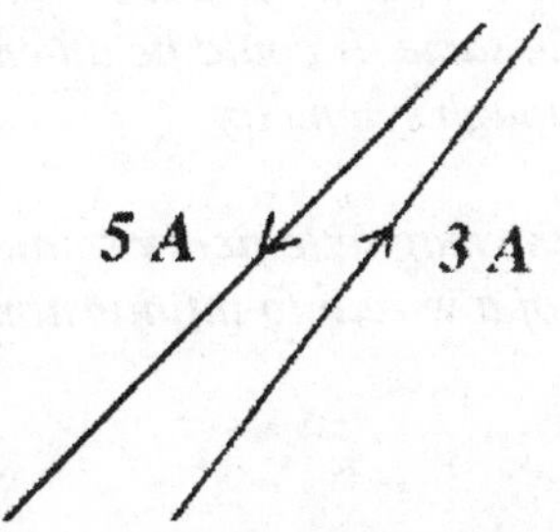

3. Two long parallel wires are separated by 0.2 meters and carry currents of 3 and 5 amps as shown in the above figure. What is the direction of the magnetic force felt by the wire on the right? By the wire on the left? How will the magnitudes of these forces compare?
Section 22-6

Most students get the direction of the force correct, but a significant minority (about 25%) get it backwards. These students often use language that shows they are comparing this situation to that of charges with parallel analagous to "like" and antiparallel analagous to "unlike" charges producing repulsive and attractive forces, respectively. Many also fall into the same mistake concerning magnitudes that they did in connection with electric forces. That is, they forget Newton's third law, and believe the larger current produces a larger force on its neighbor than vice versa.

4. Let's say you shuffle across a carpet on a dry winter day and pick up a charge of 5 microcoulombs. What force will you feel due to the magnetic field of the Earth?
Section 22-2

The primary value in this question is to establish a sense of the magnitude of magnetic forces as they compare to electric forces. In the classroom I would compare results from this question with an estimate of the force between two people each carrying the 5 microcoulomb charge. Most students will do a fine job with this question, but a few have difficulties. Most of these students are alright with the physics. Rather, the subject is sufficiently new and confusing that they balk at making all the necessary assumptions, e.g., a direction of motion with respect to the field.

5. The magnetic field can often be calculated easily by using Ampere's law. This law is similar to Gauss' Law, where we had to use an "Gaussian Surface." Now we must use an imaginary "Amperian loop." What are the essential features of an Amperian loop?
Section 22-6

In most cases, students have even more trouble with Ampere's law than they did with Gauss's law. As the responses below indicate, most have understood that the path must be closed, and many understand that it need not correspond to an object. However, the use of symmetry in selecting a loop, and in interpreting the meaning of the process is still quite shaky. For instance, many students are uncomfortable with using Ampere's law in cases where the enclosed current is zero.

Here are some sample (unedited) student responses that provide useful starting points for discussion:

a. *We use an Amperian loop that encloses some amount of current. It works for conductors of any shape, and our amperian loop can also be any shape.*

b. *The amperian loop can be any shape as long as it is a closed loop. It does not have to enclose the conductor.*

c. *The path must pass through the point of interest and it should have good symmetry. Also, the beginning and end points must be the same. It could be an empty space, embedded in a solid body, or some of each. Also, it has to have enough symmetry.*

d. *Ampere's Law is used to determine the magnetic field produced by a particular point in the enclosed surface. If we were to divide a path of a wire into infinitestimal segments then we would calculate the dot products of two scalars for each individual segments and then sum those products up. Also, Ampere's law is formulated in terms of the line instead of magnetic flux.*

e. *For example if we want to find the B for certain point the loop has go through that point. the path for is independent from the shape of the path. If B is perpendicular to all or some portion of the path, that portion of the path makes no contribution.*

6. You can make a good approximation to a "long solenoid" by buying a 50-yd. spool of wire and winding it carefully (nice, even coils) around a cylindrical core (say a broomstick), then removing the core. Estimate the maximum magnetic field that can be produced in such a solenoid.
Section 22-7

Most students find this extremely difficult. By this point in the semester, they have become comfortable estimating a single quantity, but they find it daunting to begin a problem if there are several estimations required (see responses a, b). Many students seem to ignore the phrase "long solenoid" in the question, and deal with the result for a current loop or infinite wire (c, d). In addition, many of the students that begin well fail to note that the quantity "n" is a number of turns/length rather than a pure number (e, f).

Here are some sample (unedited) student responses that provide useful starting points for discussion:

a. *We can find the field by B=uo*n*I if we were given the current.*

b. *The total current through an area bounded by the path is zero. The Amperian loop has a steady current and no time-varying electric field. Q2 = Well, if I use the equation B=MnI, I can use 4Pi*10^-7 for M, and is it okay to use 50 yards for n? I need to solve for B, But I do not know what to plug in for the current. I hope that I at least got the right equation down.*

c. *Well first I will convert 50 yds to meters which is 42.3 m take that and divide it by the circumference of each loop which would be .126 (assume .02m R.) I have to wrap around the broomstick about 362 times. Then by estimating that the current would be about 1 amp (V=120VR=140ohms) and using the b=uNI/2a I got about 1.1 X 10^-2 T.*

d. *I'm going to say the radius of a broomstick is 2cm that means that 50 yards of wire would wrap around it about 1978 times. If the current was 1 amp, then the magnetic field would be about .020 T. If the current was 2 amps, then it would be .040. That is, if I am using the right equation B=u(nought)*I*N/(2pi*r) Where N is the number of loops, r is the radius of the loops, and I is current.*

e. *Mo*n*I, where, the wire is 45.72 meters long, and the number of turns of this wire around a cylindrical core(n) is approximately equal to 1524/m. Also, assuming I=3A. Therefore, the magnetic field is equal to about 5.71x10^3 T*

f. *I will estimate that there are 900 turns around the broomstick. I will assign a current of 2 Amps. So, using the formula B=uNI, the magnetic field would be about 2.3 x 10^-3 Tesla.*

PUZZLES

"The Snake"

The next figure shows a flexible wire loop lying on a frictionless tabletop. The wire can slide around and change its shape freely, but it cannot move perpendicular to the table.
Part a: If a current were made to flow counterclockwise in the loop which of the following would happen?

a. The wire would stretch out into a circle
b. The wire would contract down to a tangled mess
c. Nothing
d. There is not enough information to tell.

Part b: If the current went clockwise, would that affect your answer to part a?
Sections 22-1 & 22-4 & 22-5 & 22-6 & 22-7

Most students get this one correct (responses e, f), but there are a variety of incorrect answers as well. Some students cling to notions about magnetic polarity that come primarily from early experiences with bar magnets. Trying to map these ideas onto problems of this sort causes a great deal of difficulty (response a). Other students have weak spatial reasoning skills, and they have trouble keeping the directions of the fields produced and the directions of several current elements straight in their heads. These students may simply build a sign error into their analysis (as in response b). It is important to help these students develop methods for keeping track of the vectors on paper (e.g., quick thumbnail sketches). Some students will conclude that the wire does not move, either because they have learned not to include "self forces (response c) or because they have learned to consider only "total forces" (response d). Both of these must be dealt with carefully, because they stem from ideals that we have taken pains to teach in previous parts of the course. The fact is, many physics courses tend to ignore flexible bodies altogether. I point this out to my students, and congratulate the class as a whole for dealing with this difficult idea successfully. There are a number of useful extensions to this problem, such as introducing an external magnetic field in either the vertical or horizontal direction.

Here are some sample (unedited) student responses that provide useful starting points for discussion:

a. *I believe that the wire would tend to contract into a tangled mess for current flowing in either direction. The reason for this is when current flows through a wire, a magnetic field is created enveloping the wire like concentric hoops. The polarity of the magnetic field is going to be opposite at points where the wire is bent and the current is flowing in "opposite" directions. We all know that opposite poles attract, which causes the wire to contract into a tangled mess. Thank you.*

b. *I believe the answer is b. If the current is traveling CW, it will create a magnetic field into the table. Then for each segment dl of the wire there will be a force perpendicular to the plane formed by dl and B. If the wire was in a circle the force would pull the wire towards the center of the circle and into a tangled mess.*

c. *I think that nothing would happen. If you ignore the magnetic field of the earth then there is nothing for the current to interact with and move into a circle or into a tangled mess. If the current went counterclockwise the results would be no different.*

d. *I think the answer is choice "c" because the force induced by a current is F=IL*B so the current cross the B field gives you force and as you add up the little bits of dF around the coil it all cancels out with each other. For b) That would not affect the outcome, it would just change the direction of the force at any point along the wire assuming there is no pre-existing magnetic field.*

e. *Part a: The wire would tend to stretch out into a circle. The current would create a magnetic field that would repel the opposite sides, where currents are traveling in the opposite direction. Part b: If current went counterclockwise, that would NOT affect my answer to part a. How? The direction of the magnetic field produced by the current might be in the other direction, but the force on one side from the field originating from the other would still create a force repelling each side.*

f. *I would say that a would be the correct answer, because as the current flows in each loop, it produces repulsive forces so that the largest forces will be repulsive ones, therefore causing the loop to expand. However, the answer to the second question would be, No, it wouldn't affect the answer from part a, because in either direction it would still expand.*

Chapter 23: Magnetic Flux and Faraday's Law of Induction

WARM-UPS

1. Lenz's law is said to be a consequence of the principle of conservation of energy. Explain this statement by describing what would happen if Lenz's law were reversed. Take the example of a wire loop with a changing magnetic flux, and describe what would happen if current were induced opposite to the way it really is.
Section 23-4

Lenz's law is more difficult for students than many instructors realize. The idea that comes through to them is "something opposes itself." They are unsure whether this means that nothing happens, or that some effect actually reverses direction. This confusion often comes out in an inability to interpret this question at all (see responses a-c). Other students have a general sense of the conservation principle, but cannot express it without reference to an accelerating object (d, e).

Here are some sample (unedited) student responses that provide useful starting points for discussion:

a. *As far as I understand, if the flux was to change direction through the loop, Lenz's law would compensate for the change. The direction of the induced current would then change.*

b. *Since Lenz's law states that the magnetic induction effect will oppose the cause of the effect, then if the current were induced to be opposite of what it really is, then the changing magnetic flux would oppose it.*

c. *If current were induced opposite to the way it is, the magnitude of the magnetic flux would change. The magnitude of the magnetic flux is dependent on the value of resistance. If current were oppositely induced, it would be equally negative in magnitude. This would mean that the resistance would either be higher or lower depending on the original sign of the current. The greater the resistance, the less the current.*

d. *Well, if the current was induced in the opposite way, then the magnetic force on the loop would accelerate to an ever-increasing speed with no external energy source. This is clearly a violation of energy conservation.*

e. *No energy would ever be dissipated. So like the book states the magnetic force on a rod would accelerate it with no external energy source present.*

f. *Lenz's law has to do with an effect that opposes the cause of an effect. If the effect went along with the cause, instead of opposing it, an ever-lasting process would be able to occur where energy comes from nowhere.*

2. Let's say you take an ordinary wire coat hanger and straighten out the hook shaped part that normally hangs over the coat rack. Now, you can spin the (roughly) triangular part around by twisting the straightened part between your fingers. Estimate the EMF that you can generate by spinning the hanger in the Earth's magnetic field (about 5×10^{-5} T).
Section 23-6

Like everything involved with Faraday's law, most students find this difficult. However, it is well worth pursuing, because the classroom discussion can be quite rich. For instance, it is a good opportunity to remind students that induction occurs due to changing flux, not due to static flux (see response a). As the responses below indicate, many students latch on to the formula for motional EMF (E = vBL) and use this to the exclusion of others (responses c, d). Discussing these responses is an ideal way to shake students loose from that formula. Finally, some students do quite well (see responses e, f). Discussing those responses helps the others in the class see how to deal with the complexities in the problem. E.g., the time dependence.

Here are some sample (unedited) student responses that provide useful starting points for discussion:

a. *The emf would be the B field times the area inside the hanger, (dB/dt*A). Assuming the hanger makes a triangle with a base of 20cm and a height of 10 cm, the area would be .02m^2, since B=5E-5, the emf would be .1E-5*

b. *I am going to assume that the length and velocity are perpendicular to B. Using equation 30-6 (emf=vBL) the emf can be calculated. I will estimate the velocity of the hanger to be 2 m/s, and the L of the hanger to be .3 m, then the emf would be 2*5e-5*.3 = 3e-5 V.*

c. *emf = vBL assuming the velocity to be about 1m/s and the length to be 0.02m and given B = 5*10^-5T, the emf = 7*10^-7V*

d. *First of all lets pick an area for the coat-hanger of about one-third of a meter squared. I would not be able to twist the hanger very fast, so, I think the db/dt would be about 5*10^-5/2seconds. The hanger is perpendicular to the Earth's magnetic field, the direction of the area vector is parallel to the Bfield. Using Faraday's law, the emf is equal to about 8.33e-6 volts.*

e. *The induced EMF equals the negative change in the magnetic flux per time. The change in magnetic flux is the change in the field times the change in the Area, 1/2bh, of the coat hanger with reference to the magnetic field. The magnetic field is constant but the Area in reference to the field goes from 1/2bh to 0 as it spins. This change is estimated per second. Therefore, if the area of the hanger loop is a triangle with the base of 15cm and height of 5cm, then the induced EMF = (5x10^-5)(1/2)(.15m)(.05m)/s = 1.875x10^-7*

f. *By estimating that I can spin a coat-hanger at an angular velocity of 5 rad/s, and that it has an area of .05 m^2. I use the formula V= wBAsin(wt) and taking the sin of 90 degrees, V=1.25E-5.*

3. Here's one way of understanding a capacitor: It is a device that won't let the voltage between two points change too rapidly, because it stores up charge and has $V = Q/C$. The charge cannot be changed instantaneously, so the voltage cannot either. Please describe an inductor in a similar way, that is, say what cannot be changed rapidly and why.

Section 23-7

Students often misunderstand what an inductor is and what it is used for. This question provides a natural opportunity to open that discussion in class. The comparison to capacitors is useful, because it can lead to a discussion of the form of the equations describing the behavior of an RL circuit, and a comparison of the solution to the solution for an RC circuit. Discussing students' responses to this question also presents a good way to bring up the role of inductors in filtering applications.

Here are some sample (unedited) student responses that provide useful starting points for discussion:

a. *Because there is so much wire involved, current cannot be changed rapidly.*

b. *The voltage across an inductor cannot change instantaneously, this is because the induced emf is proportional to the rate of change of magnetic flux, which cannot make quick changes.*

c. *An inductor is similar in the way that it cannot change rapidly because the charges don't immediately move to their positions in an inductor just like the charges don't immediately move in a capacitor. It takes time for the charges to bunch up at one end or the other.*

d. *An inductor opposes any variations in the current through the circuit. It helps maintain steady current.*

e. *This question was a bit confusing but I think I see what you want. Basically the way I understand inductions, say in a circuit is that current that is flowing could be oscillating or variations at some extents, once it comes to the inductor the inductor's job is to smooth out those variations and then pass it on through the circuit.*

f. *An inductor will not allow the current to change abruptly. The voltage across an inductor is equal to the inductance times di/dt, the change in current with respect to time. Using the equation for current in an inductor, a discontinuous, or abrupt, change in current would require an infinite voltage, which is not physically possible.*

4. Estimate the inductance of a solenoid made by buying a typical spool of wire from hardware store and winding it carefully around a broom handle.
Section 23-7

The most important feature of this problem is the clear connection it creates between the device as described in the book, and the reality of a coil of wire. Further, it continues the process of establishing a sense of scale; many students recall that a Farad is an exceptionally large capacitance and assume that a Henry is an equally outsized inductance. In the classroom, a discussion of this problem can begin with a review of the previous "broomstick problem" (Warm-Up 22-6). The classroom discussion may also emphasize the geometrical nature of inductance. Many students observe the factor of I *in the definition, and assume that the current must play a role in the inductance. Many will estimate a value for* I*, use it to calculate flux, then divide it out (see response 1). I always point out that the current will always cancel out in this way, leaving the inductance independent of* I.

Here are some sample (unedited) student responses that provide useful starting points for discussion:

a. *L = N (phiB)/I approximate N to be 200, I to be 3 A from some outlet in my house, B to be the magnetic field of the earth (5 e-5), and A to be .0004 m^2 The inductance is somewhere in the range of 1 microHenry.*

b. *If the wire is 100 meters long and the broomstick has a radius of 0.02 meters, then the number of turns equals 795 and area is 1.2 x 10^-3 meters squared. Since inductance, L, = u-nought x number turns in coil squared times area of cross section of coil divided by circumference of coil, then L = 7.59x10^-3.*

c. *I will assign some values: N=200 turns, Area=10cm^2=10x10^-4 m^2, r=0.1m. Then, I will use the formula L=((u nought)(N^2)A)/2pi(r). Placing my values into the formula-->I estimate inductance at about 80 microHenry.*

d. *L = N Phi B/ i for the self-inductance of a circuit. so L = N(nuoi)A/i = (N^2/l)(uo)(A) where N=800, l=2m, A=1.2E-3m^2 L = 4.83E-4.*

5. A wire hoop surrounds a long solenoid. If the current in the solenoid increases, will there be any current induced in the loop? If so, which way will the current flow?
Section 23-6

Some students will make the classic mistake of believing that the induction only occurs if the magnetic field is "touching" the conductor, as in response a, below. In fact, many students that do not make the mistake do so not out of deep understanding, but because they forget that the solenoid produces no external field. In the classroom, I will often discuss the field of the solenoid first, then take a quick "poll" of the class to see if anyone's answer is affected. Only after discussing this will I move on to the issue of direction (with current increasing and decreasing). Many students that are correct on the issue of whether induction occurs are shaky about the direction of the induce current (see response b).

Here are some sample (unedited) student responses that provide useful starting points for discussion:

a. *No, there will not be any induced current. There is no B field outside the solenoid.*

b. *Yes there will be current induced in the loop. The changing current will produce a changing B field. A changing B field produces a changing flux. The changing flux about the wire will produce a current in the downward direction.*

c. *Yes, a current will be induced because a changing magnetic field will cause an electromotive force which causes there to be a current. The current will flow in a direction that would oppose the change. It will flow in the direction opposite if the current increases. So if it is flowing clockwise, then the induced current will flow counterclockwise.*

d. *The increase in current in the solenoid will increase the magnetic flux as well. This increase in magnetic flux will cause a induced charge in the loop which will obey Lenz's Law concerning direction. That is to say that the induced current will create a magnetic field in an opposite direction to the field that created the current. In this case the new B field will point to the left in the picture and the current will flow clockwise around the loop if looking at a cross-section facing the end marked I.*

6. A car battery has an emf of only 12 V, yet energy from the battery provides the 20,000 V spark that ignites the gasoline. How is this possible?
Sections 23-9 & 23-10

This question can be tricky to discuss in class, because of the wide range of knowledge students have about cars. This is an opportunity for a confidence builder for students that are struggling in class but know a lot about cars, but it is important to remember not to offend students that have no automotive experience. The responses below show this wide range of foreknowledge. Responses a and b come from students that may have heard the names of a few car parts, but are completely unaware of when and where the "sparks" occur. In contrast, responses d-f come from students that know ignition coils are essentially transformers. Response c comes from a student that has an understanding of what an inductor might do, but does not refer to the car at all.

Here are some sample (unedited) student responses that provide useful starting points for discussion:

a. *The solenoid acts as a conductor which can produce a self-induced emf. This creates a large voltage increase over the circuit when the key to the car is turned on to close the circuit.*

b. *The emf is an average energy, and can crank over the 20,000 V spark for the car. While the car runs, it recharges the battery so that it will be able to do this again.*

c. *V sub L = -L*di/dt. If either L or di/dt is very large this would be possible to create such an effect.*

d. *A car has an ignition coil that consists of a mutual inductance type relationship between the battery emf and the ignition emf. The differing emf is because of the relationship between the coil sizes and respective currents.*

e. *The ignition coil provides the high voltage from the energy that is stored within its electric field coils. The high voltage that is developed in one winding of the car ignition coil results from the sudden flux change caused by interruption of the current in another winding.*

f. *The ignition coil provides the high voltage from the energy that is stored within its magnetic field coils. The high voltage that is developed in one winding of the car ignition coil results from the sudden flux change caused by interruption of the current in another winding.*

PUZZLES

"Play Bulb!"

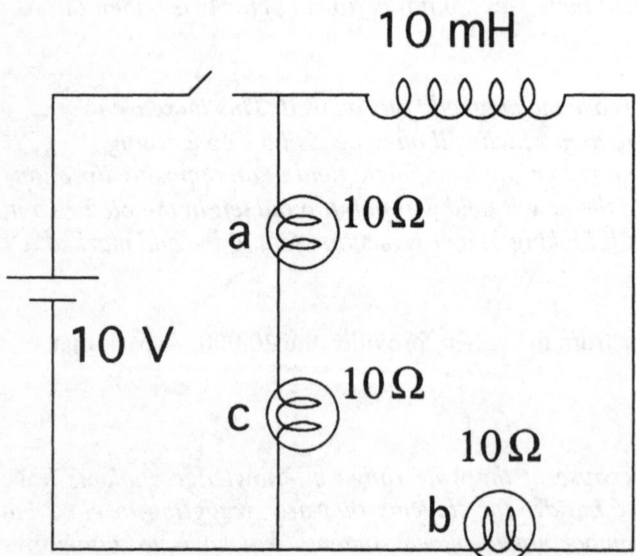

The above figure shows three identical lightbulbs attached to an ideal battery and an inductor. The switch is closed for a long time, then opened. Which of the following statements correctly describes what happens just after the switch is opened?

a. Bulb B stays the same, and bulbs A and C go out.
b. Bulb B stays the same and bulbs A and C get brighter.
c. Bulb B goes out, and bulbs A and C get brighter.
d. Bulb B gets brighter and bulbs A and C stay the same.
e. Bulb B gets brighter and bulbs A and C go out.
f. All three bulbs get brighter.
g. All three bulbs go out.
h. None of the above. (State what DOES happen)
<u>Sections 23-7 & 23-8 & 23-9</u>

This puzzle shows up many students that have learned some aspects of dealing with inductors, but have not internalized all of the important aspects. For instance, see response a, in which the student has heard the message that "an inductor acts like a short circuit after a long time" but has not understood that energy storage goes on as a result. Others have understood that the inductor "acts like a source" after a change in the switch, but are not sure what kind of source or how to work with that statement (b and c). Many students do fine with this puzzle, though, as in responses d and e. There are many ways to extend the discussion of this puzzle: What if the inductor were switched to the other branch? What if the inductor is replaced with a capacitor? Can capacitors be arranged to have a similar effect? What happens if the inductor is in series with the switch? Is that ever a good idea?

Here are some sample (unedited) student responses that provide useful starting points for discussion:

a. *Answer g, all three bulbs go out because after the switch closed for a long time, an inductor became short circuit, none of energy was stored inside an inductor. If the switch is opened, there is no current flowing through bulbs A, B or C. And all H. B remains the same and A and C get dimmer. This is because B goes through the inductor both before and after the switch is thrown.*

b. *The answer is h. All three bulbs get dimmer. Just after the switch is opened, the inductor acts as an emf source equal to the battery because of continuity. Before, the current through B is 1 A and .5 A through A & C. After, the current through all three is 1/3 A.*

c. *All three bulbs get brighter. When it has been close for long time, there is no current in B(no volt), but in A and C current are 0.5A. Therefore, each bulb has 5 volt When switch just opened, current in A, B, and C are the same 1 volt. So all three bulbs get brighter.*

c. *Before the switch is opened, the inductor is a short. This means bulb B gets 1 amp of current and bulbs A and C get .5 amps. After the switch is opened, the inductor acts as a current source. The three bulbs are in series. This means all three bulbs have the same current. The current the three bulbs have is 1 amp. This means bulb B stays the same and bulbs A and C get brighter. Answer (b).*

e. *The current through the inductor, after the current is constant, is 1 A (by V/'B'). When the switch is opened, that current remains the same for an instant. However, now A, B, and C are in series and all experience the same current. Before, A and C had current equal to 0.5 A (by V/('A+C')). Since the brightness is determined by P=I^2R, B will remain the same for an instant while A and C will get brighter for that same instant.*

Chapter 24: Alternating-Current Circuits

WARM-UPS

1. Home Experiment: Make a pendulum using a string 1 to 1.5 meters long (40–60 inches) and a mass similar to a lemon (go ahead, use a lemon!). This will give you a pendulum with a resonant period in the range 2 to 2.5 seconds ($T = 2(L/g)^{1/2}$). In terms of frequency, $f = 1/T$, so the resonant frequency is between 0.5 and 0.4 Hz. Now, hold the end of the string and vibrate it back and forth by 5 cm, slowly varying the frequency of your oscillation. When you find the right frequency, the lemon will swing back and forth with quite a large amplitude. This is the resonant period for your pendulum. Note the phase between the motion of your hand and the lemon. (Is your hand out ahead of the lemon pulling it forward, or behind pulling it back?) Try swinging the lemon a little more slowly (longer period than at resonance), how does the amplitude change? How about the phase? Next, try going a little more quickly than at resonance. What happens to the amplitude and phase now? Finally, compare the experiment you just did to the behavior of an RLC circuit.
Section 24-6

Most students find this type of reasoning difficult, even if they have fun with the experiment. Instead of asking only about the source, I used to ask for analogies to the inductor, capacitor, current amplitude, etc., but I abandoned those as being too difficult. However, I still make this discussion an important part of the classroom activity, after demonstrating the experiment.

Here are some sample (unedited) student responses that provide useful starting points for discussion:

a. *Amplitude increases when we try going slower than at resonance and decreases when we go faster than at resonance. I'd have to guess and say that the amplitude is the equivalent of the AC source.*

b. *When swinging the lemon in the experiment slowly the amplitude decreases, and going a little more quickly than at resonance did not produce a larger amplitude than at resonance. This seems to correspond with the definition of resonance: The peaking of the current amplitude at a certain frequency. The maximum value of I occurs at the frequency at which the impedance Z is minimum. Comparing this experiment to a RLC circuit, the ac source is the lemon, and the oscillations in the pendulum movement is the angular frequency.*

c. *When I swung my "lemon" more quickly, the amplitude decreased; when I swung it more slowly, the amplitude decreased and the "lemon" became difficult to control, as it would hit the trough and jerk the string. I think the ac source is similar to my movement of about 5cm back and forth. At the point where the amplitude was maximum, the "lemon" was in resonance.*

d. *Not having a lemon around I had to use other fruits and vegetables, and also a shorter string had to be used, for some reason I kept getting the (apple) to spin in a sort of circular motion, I don't think that this was what was intended...I will, however, say that the amplitude of the fruit will go down, when you either increase or decrease the frequency of the source your hand is the source of the whole thing, so that would be the ac source then.*

e. *By moving a little more quickly than at resonance the amplitude decreases. Basically the faster you go after the resonance the smaller the amplitude. My hand and is the AC source. Comparing this with an RLC, I would say that as you increase the angular frequency slowly you come to a point where the current amplitude is the largest amplitude and the resistance is smaller that at the beginning, but as the angular frequency increases amplitude of the current decreases and the resistance increases.*

2. A radio has RLC circuits inside that oscillate at the same frequencies as the radio waves they receive (88.1-107.9 MHz for FM, 540-1180 KHz for AM). Select values of L and C that could be used to make these circuits.
Section 24-6

I am always amazed by the difficulty this problem presents to my students. The physics is easy, and if I give them L, *they can calculate* C *for resonance with no problem. However, they are extremely uncomfortable with the ambiguity this problem presents. Many students simply balk (responses a, b). Others calculate values for LC and then quit. Less than half of the class will go ahead and choose a value for one parameter then calculate the other. In most cases, these are students who have electronics experience either on the job or as a hobby.*

Here are some sample (unedited) student responses that provide useful starting points for discussion:

a. *For problem two I must have had a mental block because I tried to figure it out for a half-hour. I didn't know if X(l) = X(C) was the right equation to use for the problem or not. I just didn't know where to start. The only thing that I could find would be the angular frequency. This is a tough problem for me.*

b. *I could use an equation, w=[1/(sqrt(LC))], but then I am stuck with two unknowns that I am solving for!*

c. *Since the angular frequency, w = 2 x pi x frequency, then the angular frequency ranges for AM and FM would be: w(AM) 3.39 x10^6 to 7.4 x 10^6 w(FM) 5.5 x 10^8 to 6.8 x 10^8 so LC=(1/w)^2 and the ranges for LC would be: LC(FM) 3.3x10^-14 to 2.17x10^-18 LC(AM) 8.7x10^-14 to 1.8x10^-14*

d. *f=w/2pi for frequency 99.5MHz w=625e7Hz w=1/sqrtLC LC=2056E-20 any combination which would give above number should probably work.*

e. *Use the equation omega=1/(LC)^(1/2). For FM, omega can be 553-678, and for AM it can be 3393-7414. This would give for FM that LC = .0018-.0014. For AM it would be .00029-.00013. Practical numbers for C would be 8e-6 F for both FM and AM, and then L would be around 200 H for FM and 26 for AM.*

f. *Well, we know angular frequency equals 2*pi times frequency which equals 1/sqrt(L*C). By selecting an appropriate value for the inductance, we can find the range of the capacitance needed to reach these frequencies. For AM stations, we will choose a 1 micro-henry inductor. This will give us a capacitance range needed from 9.9 nF to 86.9nF. For FM stations, we will choose a 4 micro-henry inductor coil. This will give us a capacitance range needed from .543 pF to .818 pF.*

3. An inductor or a capacitor in an ac circuit has a (time varying) current flowing through it and voltage across it. However, the average power $P_{AV} = 0$. Explain how this works out, recalling that power is equal to voltage times current.
Section 24-4

This usually breaks down according to the students that gave the question some thought, and those that leaped to a conclusion. The latter category are those that give answers similar to a and b, below. I usually start the discussion of this question by showing an answer of this type, praising it as a good start, but then analyzing the power in the resistor. I then discuss the correct answers, and use the whole thing as a bridge to discussing the power at the ac source as the intermediate case (phase between 0 and ± $\pi/2$).

Here are some sample (unedited) student responses that provide useful starting points for discussion:

a. *This is very simple to explain. For each positive voltage there is an equal but negative voltage. So for each pint in the cycle there is an equal and opposite voltage later in the cycle to cancel it. Your voltage average is zero.*

b. *I believe that the reason the average power is zero lies in the changing current. If the current is alternating in a sinusoidal fashion, the AVERAGE current will be zero, therefore making the product of current and voltage zero.*

c. *In the inductor what is happening is that the voltage leads the current by 90 degrees. When we take the product of the voltage times the varying current, and we graph it we see that the power graph is positive half of the time, and negative the other half. This leads to an average power of zero in the inductor.*

d. *A capacitor or an inductor current is not in phase with the voltage in an ac circuit. It is what we say 'out of phase' (leading or lacking by 90 degrees). This means when the current is max, the voltage is 0 and vice versa.*

e. *Energy is supplied to the component during one half of an AC cycle. The energy which was stored in the component is discharged when the current changes direction. It is used to charge the source, therefore the total energy in one full cycle is zero.*

4. Old radios used to tune from one frequency to another using variable capacitors like the one shown here. In such a device there are two rows of semicircular plates. One set of plates (all electrically connected together) are fixed to the base, and the other set (all connected together) can rotate through a half turn. At one extreme they are opposite to the fixed set, and at the other extreme they are fully "interleaved" with the fixed set. Assuming the whole thing must fit inside an old radio, estimate the maximum capacitance. (Hint: for 10 fixed and 10 rotating plates, you can treat this as 10 parallel plate capacitors in parallel.)
<u>Section 24-3</u>

This is a tough question, but it is worth assigning occasionally. Students can usually deal with calculating the capacitance of the thing if they are given the dimensions of the plate pairs, but they have a tough time envisioning the whole assembly fitting into a radio, and coming up with the dimensions on their own.

PUZZLES

"Ups and Downs"

The below figure shows a graph of the current in a series RLC circuit, $I=I_{max}\cos(\omega t)$. There are five points marked on the graph, labeled A, B, C, D, E. Please state at which point (or points) each of the following quantities are a maximum, a minimum, or zero:

a. The energy stored in the inductor
b. The voltage across the capacitor
c. The power input to the capacitor
d. The power output by the inductor
e. The charge on the capacitor
f. The voltage across the resistor

Sections 24-1 & 24-2 & 24-4 & 24-5

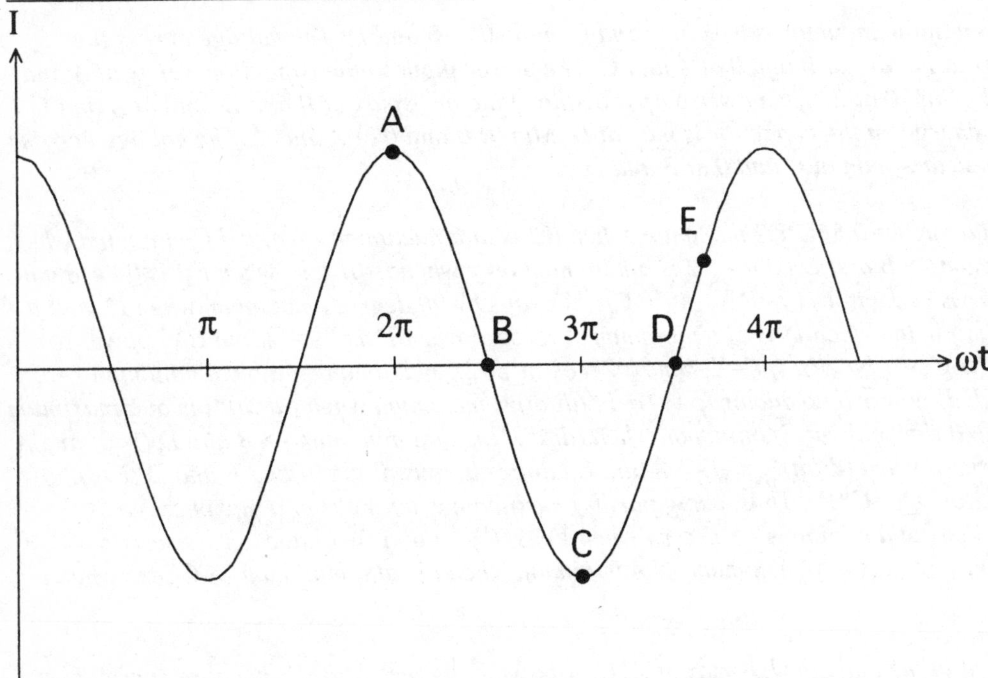

This is a relatively difficult puzzle. It requires a fairly complete understanding of a difficult subject. It is also long, and challenges many students' organizational abilities. In particular, many students have a great deal of difficulty with parts c and d. I usually give partial credit for this puzzle (I do not for most) because so few students get the entire thing correct. Extensions to this problem include anything from more of the same (do the same for the voltage across the inductor and the power dissipated in the resistor), to complete reworking (let this be a voltage graph, now redo the problem). I have listed several responses below to give a flavor of the range of answers that my students submit.

Here are some sample (unedited) student responses that provide useful starting points for discussion:

a. *The energy stored in the inductor is a maximum when the current is a maximum, negative or positive... doesn't matter. The voltage across the capacitor is a maximum when the current is going through zero. The real power input to the capacitor, being reactive, is all the time zero. The power output by the inductor is that which is burned by the resistor and is maximum when the current is maximum. The charge on the capacitor is full when the voltage is full, that is when the current is going to zero. The voltage through the resistor is maximum when the current is maximum. For the minimums, we should be able to use the converse and a lot of the minimums will be "zero"*

b. *The energy stored in the conductor is at its maximum at point A, minimum at point E and zero at points B and D. The voltage across the capacitor is at its maximum at point B, minimum at point D, and zero at points A & C, due to the voltage formula and its relationship to I. The power input to the capacitor is at its maximum at point B & E, zero at A, and minimum at D. The power input by the inductor is at minimum at B, zero at D, and maximum at C because the wavelength of the power is virtually half that of the current. The voltage across the resistor is at maximum at A, zero at D and B, and minimum at E. The charge on the capacitor is zero at A and C, minimum at D, and max at B. The answers are based on different graphs of voltage, current, and power for various circuits shown in the book.*

c. *The energy stored in the inductor is max at A and C, min=0 at B and D. The voltage across the capacitor is max at D, Min at B and 0 at A and C. The power input to the capacitor is max at A and C, min at B and D and 0 at E. The power output by the inductor is max at B and D, min at A and C and 0 at E. The charge on the capacitor is max at D, Min at B and 0 at A and C. The voltage across the resistor is max at A, min at C and 0 at B and D.*

d. *a) energy in inductor: w=0.5*L*i^2 maximum when i^2 is at a maximum -> A and C minimum when i^2 is at a minimum -> B and D (since i^2 is a minimum only when i=0) zero when i^2 is 0 -> B and D b) voltage across capacitor: Vc=I*Xc=i/(o*C); *Vc lags by 90 degrees maximum when I* is at a maximum -> B minimum when I* is at a minimum -> D zero when I* is 0 -> A and C c) power to capacitor: p=v*i=C*v*(dvc/dt); (pc=Xc*Imax^2*cos(ot-pi/2)) maximum -> a,b,c,d minimum -> e zero 0 -> a,b,c,d. d) power by inductor: p=v*i=L*(di/dt)*i maximum when (di/dt)*i is at a maximum -> E (di/dt!=0 and i!=0 only at E) minimum when (di/dt)*i is at a minimum -> B and D (i=0) and A and C (di/dt=0) zero when (di/dt)*i is 0 -> A and C (slope=0, thus di/dt=0) and B and D (i=0). e) charge in capacitor: Qc=C*Vc. Thus, using part b) maximum when Vc is at a maximum -> B minimum when Vc is at a minimum -> D zero when Vc is 0 -> A and f) voltage across resistor: Vr=i*r maximum when i is at a maximum -> A minimum when i is at a minimum -> C zero when i is 0 -> B and D*

e. *The energy stored in the inductor has max at A, C; Min which is zero at B, D; because energy U = 1/2*L*I^2 = 1/2*L*I(max)^2*(coswt)^2, from the equation, it never be negative, so the minimum value is zero. The voltage across the capacitor has max at B, min at D, zero at A,C, because of the Vc=Xc*I(max)*cos(wt-Pi/2), the voltage of the capacitor lags the current by 90 degree. The power input of the capacitor has min at E, zero at A,B,C,D. Because the power for the capacitor Pc=Vc*I = Xc*I(max)^2*cos(wt)*cos(wt-Pi/2). The power output by the inductor has max at E, zero at A,B,C,D. Because the power for the inductor PL=VL*I = XL*I(max)^2*cos(wt)*cos(wt+Pi/2). The voltage in inductor is leading the current by 90 degree. The charge on the capacitor has max at B, min at D, zero at A,C. Because Qc=C*Vc = C*Xc*I(max)cos(wt-Pi/2). The voltage across the resistor has max at A, min at C, zero at B,D. because of VR= I*R=R*I(max)cos(wt).*

f. *Energy stored in the Inductor is maximum when the current is a maximum. U=1/2*L*I^2: Thus; Point A: max Point B: zero Point C: max Point D: zero Point E: min Voltage across the Capacitor: Since the Vc lags the current by 90deg the voltage across the capacitor will be maximum when current is zero and will be zero when the current is maximum. Thus; Point A: zero Point B: max Point C: zero Point D: max Point E: min The power input to the capacitor has 2 cycles for 1 current cycle* *and since the power is zero when the current is zero this means that the power curve crosses the axis(zero) when the current is also at maximum. Thus; Point A: zero Point B: zero Point C: zero Point D: zero Point E: min The power output by the inductor is P=Ldi/dt=Vab*i. This implies that since the VL leads I by 90 deg then power is maximum at increments of 45 deg and is zero when current is maximum. Thus; Point A: zero Point B: zero Point C: zero Point D: zero Point E: max The charge on the capacitor is q=C*Vc. This implies that the charge is maximum when the voltage across the capacitor is maximum. Thus; Point A: zero Point B: max Point C: zero Point D: max Point E: min The voltage across the resistor (Vr=I*R) is in phase with the current so where current is maximum the voltage across the resistor will be maximum and the same for zero and minimum values. Thus; Point A: max Point B: zero Point C: max Point D: zero Point E: min.*

Chapter 25: Electromagnetic Waves

WARM-UPS

1. In a wave on a string, the amplitude is the maximum displacement of a point on the string from its equilibrium position. What is the amplitude of an electromagnetic wave?
Section 25-1

This is a great question for turning up students' misconceptions about what is going on at a fundamental level. Many faculty take for granted that the students understand that an electromagnetic wave is composed of E and B fields, then dive right into the mathematics of plane waves. In fact, many students are deeply confused about fundamental issues such as "what is it that has a maximum," and "what is moving at c." See responses a-c for examples of this confusion. Furthermore, the static diagrams of plane waves in our texts tend to exacerbate this confusion rather than helping to clear it up, responses c and d, below, show that students are still trying to interpret the "wave" as a displacement of something rather than as a graph of effect vs. location. Other confusions are also common: answer f comes from a student that has (possibly) got the right ideas about the fields, but is confused about the phases.

Here are some sample (unedited) student responses that provide useful starting points for discussion:

a. *I can't find ANYTHING that will isolate the A for me. The only equation I keep finding is Asin(omegat-kx)*

b. *The amplitude of an electromagnetic wave depends on what type of field created it. If a magnetic field created it then the amplitude is the maximum of the B field. If it is created by a electric field then it the maximum value of the E field.*

c. *It is the maximum amount of disturbance from the normal route without the disturbance.*

d. *Amplitude always means amplitude. So the amplitude of an electromagnetic wave is the maximum displacement from a point on the wave (remember though that the wave is not made up of particles, so this point isn't really a point?) to its equilibrium position.*

e. *I would have to say that the amplitude = 0. Electromagnetic waves (unlike mechanical waves) need no medium to be transmitted. Only electric and magnetic fields are 'waving' in an electromagnetic wave, not the wave itself.*

f. *The amplitude of an electromagnetic wave varies per instant in time. at B max E zero and vice versa. It would appear that as a combination of B and E that the combinations of the E and B oscillations would prevent a true amplitude like one on a sine wave.*

g. *An electromagnetic wave is comprised of both an E-field and a B-field that sustain each other, so an electromagnetic wave is given by two equations, one for the B-field and one for the E-field. Thus the amplitudes of these equations are E-max and B-max, the maximum values of the E and B fields that make up this wave.*

2. Estimate the wavelength of your favorite radio station.
Section 25-3

This is a fairly straightforward question, and most students get it right (as in responses b, c). The majority of incorrect responses have to do with confusion between Hz, kHz, and MHz, as in response a. Nevertheless, I continue to ask this question because of the level of involvement it produces. Most students are devoted to their music, and the responses to this question show it. I have occasionally shown "statistics" in class on local radio preferences.

Here are some sample (unedited) student responses that provide useful starting points for discussion:

a. *Taking that my favorite station is 103.3kHz, the wavelength will equal c/f or 2904 meters*

b. *The wavelength of my favorite radio station can come from the equation lambda=c/f, where c is the speed of light. THE BEST! radio station is 98.3 which is .983*10^8Hz, so once plugged into the equation gives a wavelength of 3.052m.*

c. *Remembering c=3*10^8 and the frequency of my favorite station to be 98.3MHz (WXIR-Great Christian Music-Check it out), and c=hf (pretend the h is a lambda), the wavelength is about 3m, assuming my calculator worked right.*

d. *My favorite is WAMU 88.5 FM (which means 88.5MHz) in Washington DC. l=c/fl=(3.0 X 10^8)/(88.5 X 10^6) l=3.389 meters*

3. At a given distance, say, 3 m, the light from a 200 W lightbulb is brighter than the light from a 60 W lightbulb. Does this mean that light coming from the 200 W bulb has a larger electric field? A larger magnetic field? A higher frequency? A longer wavelength?
Section 25-4

This is a relatively difficult question. Most students determine that E is bigger, but many ignore the effect this has on B. Some students know (at least roughly) about the relationship between frequency and photon energy, particularly those that had a physics course in high school or a university level chemistry course. In this case, they may also believe that the frequency increases/wavelength decreases for higher power bulbs. In class, a discussion off this question provides a natural starting point for a discussion of intensity and the Poynting vector.

4. Light from the Sun reaches the Earth's atmosphere at a rate of about 1350 W/m^2 Assuming that this light is entirely absorbed, estimate the force exerted on the Earth due to radiation pressure. Is this significant compared to the force on Earth due to the Sun's gravitational attraction?
Section 25-4

This is a relatively easy question; students that have read the chapter will generally be able to do it. It makes a good starting point for discussions of more detailed issues such as energy density and the strengths of the electric and magnetic fields. Students should also be asked to do a similar calculation for a common lightbulb, and to consider the variation of all of these quantities with distance. In particular, it is valuable to work through the variation of, e.g., the electric field strength as a function of distance from a spherically symmetric source.

PUZZLES

"Light Drills"

The sketch below shows the basic elements used by A. Michelson to measure the speed of light in the 1920's. A pulse of light hits the rotating 8-sided mirror and takes off on a round trip to a flat mirror some distance off. The pulse will be received by the detector if the rotating mirror is in the right orientation when the pulse returns. A moderate price power drill can turn at about 1200 RPM. If such a drill is used to turn the mirror, how far away does the flat mirror have to be for a pulse to be observed? How about for a Dremel that turns at 30,000 RPM? (Hint: Assume the distance to the mirror (L) is large compared to the size of the 8sided mirror.)
Section 25-1

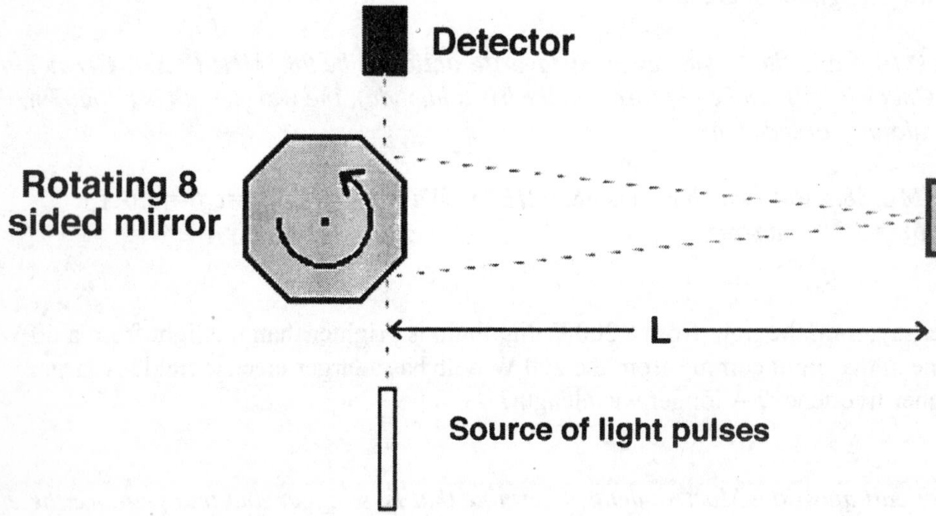

Detector

Rotating 8 sided mirror

L

Source of light pulses

This is not too hard a puzzle, and most students will either get it outright, or they will ignore the symmetry of the mirror and get a length too big by a factor of eight. Indeed, many students probably could have done this puzzle in the first few weeks of the mechanics class, as d = vt is really all of the physics in this problem. Nevertheless, students enjoy this puzzle as a pure intellectual exercise. Most students are also impressed at how large the distance must be (Michelson did the experiment in the Rocky Mountains, with the drum and the mirror on separate peaks). It is a good puzzle to bring a bit of life to what is often a dull part of the course. The best extensions of this discussion focus on the experimental and historical aspects of measuring the speed of light. It is particularly good if you have a lab in which students measure c using modern methods.

Here are some sample (unedited) student responses that provide useful starting points for discussion:

a. *Here goes, I hope you're up to reading a bit. When light hits the mirror the first time, it has to hit flat mirror and then the octagonal mirror again. The octagonal mirror must be in the same position as it was the first time for the light to be sensed by the detector. This means that it has to have moved at least one eighth of a revolution or in some multiple of an eighth of a revolution. If the octagonal mirror moves only one eighth of a rev this is the best case scenario since it would take the least amount of time. We will assume that by the time that the beam of light returns to the mirror that the mirror has turned only an eighth of a revolution. For a 1200 RPM drill this would only take 0.00625 seconds. With this amount of time, the light could travel 936.875 km twice so this would be L. For the 30,000 RPM drill it would only take 2.5 x 10^-4 seconds to turn one eighth of a revolution. This would give the light beam time to travel 37.475 km twice which would be L. I'm quite proud of myself for coming up with such an answer even if by some chance it is wrong.*

b. *This must have been a tough experiment to carry out! My distances are definitely very large. 1200 RPM is 20 RPS and the mirror only has to go 1/8 of the way around to be back to the proper orientation which at 20 RPS will take .00625 seconds. I simply used d = vt with x = 2L, v = c (speed of light) and t = .00625. Solving for L I get a distance of 937.5 km! Wow! For 30,000 RPM I did the exact same thing but with the mirror spinning so much faster the time for it to go 1/8 rev was much smaller (2.5 E -4 sec) so the distance was shorter, 37.5 km, which is better but still awfully far!*

c. *Assuming that the light travels from the 8-sided mirror to the flat mirror and back in 1/8 of rotation of the 8-sided mirror. Then, the flat mirror would have to be 937500 meter away using speed of light = 3e8. First the rpm is turned to rad/s. 1200 rpm = 40*pi/s. Multiply this with the rotation of the mirror (1/8*2*pi)1/4*pi to get the time it takes for the mirror to turn this much. Finally set the time it takes the mirror to turn equal to 2L/c where c is the speed of light. We find that L would have to be 937500 meters away for the light pulse to be detected. The Dremel that turns at 30000rpm would require the flat mirror be placed 37500 meters away. That's a very big distance. I think something might have gone wrong with my formula somewhere.*

Chapter 26: Geometrical Optics

WARM-UPS

1. In your own words, explain what a focal length is. Try not to use any equations or refer to any specific type of mirror or lens.
Section 26-3

It is tempting to think that students would have no preconceived notions about the term "focal point." The responses below indicate that this is not the case. I use a discussion of this question to emphasize that the focal point (and length) are completely fixed for a given lens or mirror, and do not depend either on the location of the object or the actions of the observer. In fact, I find it important to explicitly discus what an observer sees and does throughout the treatment of geometrical optics.

Here are some sample (unedited) student responses that provide useful starting points for discussion:

a. *A focal point is a point in space where several lines converge.*

b. *The focal point is the point that is focused on when looking at an object in a mirror. It is where the object looks to be located at.*

c. *A focal point is a point that you pick to look at. For example, you are looking at a picture and you pick to look at an apple. However, someone walks by and stands right where the apple is. If you want to keep looking at the apple you have to see it, because the person is blocking the light rays of the apple. You need to find a place where the rays are not blocked to have a view of the apple.*

d. *A focal point is where light waves come together to form an image that is clearest. It is where the light waves that are reflected off the object come together to form the correct image of the object. You can see the image at different distances away, but only at the focal point is where the image is the clearest.*

e. *I think the focal point is the part of the lens that the object is most magnified. Or the concavity is the greatest, which means everything will go out of focus.*

f. *A focal point is the point in space where any rays that come in parallel to the optic axis will reflect off the mirror and pass through it. Also, if an incoming ray passes through the focal point, it will be reflected parallel to the optic axis.*

2. Estimate the focal length of a typical bathroom "magnifying mirror."
Section 26-4

This is a very difficult question for most students. A common mistake (which I never imagined before using this question) is for students to imagine a round mirror and use the radius of the mirror in place of the radius of curvature of the mirror (see responses a, b). Students also commonly fold the confusion between focal point and "point of focus" into their answers to this question (see Question 26-1 responses b, c and responses c, d here). While some students recognize that the radius of curvature must be large, very few cite their experience with such mirrors. Optionally, this question can include a description of what one sees at varying distances from the mirror.

Here are some sample (unedited) student responses that provide useful starting points for discussion:

a. *If the R or radius is 10cm then using f=R/2, the focal length would be .05 meters.*

b. *I would have to say that the focal length for a makeup mirror would be about 1/6 of a meter because I would guess the radius of a makeup mirror would be about 1/3 of a meter.*

c. *If the object is .3m away from the mirror then the distance is R/2. Then the radius is .6. This is also the focal length because the image distance can be as far away as you want it to be.*

d. *The mirror I have in the bathroom is about 3 feet away from where I stand, so I would estimate that the focal length is 3 feet away from the mirror.*

e. *It seemed my image really began to get magnified at about 7 cm from the mirror so the focal length is about 8 cm.*

f. *I would say since the mirror is usually about 5inches across and has a barely perceivable convex curve to it, the radius would probably be kind of large. So if I use 36 inches as my radius and divide it by 2 that gives me focal length of about 18 inches or 27 cm.*

3. Images formed by spherical lenses can be real or virtual. What does it mean for an image to be virtual? Can such an image be seen?
Section 26-7

Students are often confused by the notion of a virtual image. As the responses below show, many believe that virtual images cannot be seen (responses a-c) or which can be seen but only in some artificial sense (responses d-f). About a third of the class seems to really recognize that a virtual image "looks" just as real as a real image. I usually stress, in a discussion of this question, that in fact we all see and use virtual images routinely, in common mirrors and through eyeglasses and contact lenses. (I usually do not bring up the interpretation of things seen through a flat window as virtual images except with particularly good students.)

Here are some sample (unedited) student responses that provide useful starting points for discussion:

a. *We can see Real Images and not Virtual Images. An image is Virtual if its light rays do not pass through the image point. Real Images have light rays that pass through the image point.*

b. *A virtual image has light rays that do not pass through its image point, where a real image's light rays do pass through its image point. Real images can be seen but virtual images cannot.*

c. *A virtual image is one that is constructed by the lens. It can be seen only while looking through that lens.*

d. *A virtual image can be seen, but it doesn't actually exist. It can only be seen when looking into a lens or mirror.*

e. *A virtual image is one wherein images of the outgoing rays don't pass through the image. Yes it can be seen with the correct equipment.*

f. A virtual image is when reflected light rays appear to be coming from the image point but are not actually passing through the point. They can indeed be seen.

g. If the outgoing rays don't actually pass through the image part, we call the image a virtual image. Yes the virtual image can be seen.

4. Estimate the focal length of a typical magnifying glass.
Section 26-7

This is another good opportunity to give students a feel for the "typical" numbers, and to encourage experimentation with a fairly common household device. Most students will attempt to make the estimate by gauging the image and object distances, but a few will refer to using a magnifying glass to burn paper, etc. in bright sunlight. Discussing this process also provides an occasion to bring in the size (diameter or area) of the lens. Many students have a sense that this parameter is important, and are continually frustrated that it plays so weak a role in our discussion of lenses.

5. What is the focal length of a clear, flat, piece of glass, such as a normal household window?
Section 26-7

The results with the question, and the value to discussing it, are closely aligned with the similar question concerning plane mirrors (Question 26-3). While many students are confused by the question, it gives ample opportunities to discuss the physics of thin lenses, as well as providing a good example of "thinking like a physicist" by treating this to be a limiting case of both the converging and the diverging lens.

PUZZLES

"Physics Lite"

The figure below shows a concave mirror with $f_M = 12.5$ cm and a converging lens with $f_L = 25$ cm. They are placed 50 cm apart with an object centered between them. The lens will create images of both the object and its reflection. Describe the size and location of the images formed. Bonus question: Describe how this setup could be useful in a lighthouse.
Sections 26-3 & 26-4 & 26-6 & 26-7

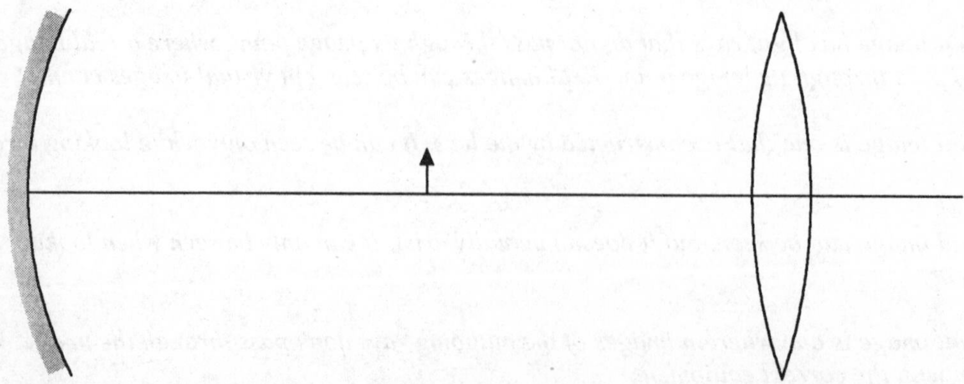

Most students easily figure out what the mirror does here, but at least half have great difficulty with the lens. Most students are uncomfortable with "infinity." They are used to eliminating it ("you can't divide by zero") or excusing it ("acceleration won't really be zero because there really are no massless ropes") but here is a problem in which infinity is the correct answer. As a result, even students that calculate the answer correctly often give incorrect interpretations. In the classroom, I use this problem to begin a conversation about a range of "infinities" such as "Can you focus your eyes at infinity? Can you see something at infinity? Does infinite distance depend on a particular lens? How? Students need to understand the answers to these questions before the treatment of most optical instruments can make sense.

Here are some sample (unedited) student responses that provide useful starting points for discussion:

a. *The lens would create a inverted real image, by itself. So the image of the mirror and lens combination would make multiple images, real and virtual. A real image would be at infinity, and a virtual image would be behind the mirror. The lighthouse can direct light out into infinity, and yet have a direct point for ships to see, and be aware of the rocks or cliffs.*

b. *Basically, the object is reflected off the mirror and you get a "virtual" image of the original right below the object. The lens then sees the object and the virtual object at its own focal length. Since the object and its mirror image are right on the focal length, the s'(image distance) is at infinity so you could look into the lens from any distance and see the image.(it never focuses) Actually, it would be blurry since it never focuses.*

c. *First, the object's light will reflect off the mirror on the left and make an inverted, exactly same size of image at the same position the object is at. Next, these two (image and object) projections will go through the lens but the image will never really finely appear. Since the image is at the focal point of the lens, the principal rays will never converge. It will just make a blurry light with not definite shape.*

d. *The mirror produces an image at the focal point of the lens and it's upside down. Also the lens produces an image of the arrow at infinity. The lens also produces a image of the image that the mirror made at the lens focal point, and puts it at infinity also. This principal is used in lighthouses to produce a beam of light twice as wide as what is producing the light by putting a light source that shines on the mirror and the lens at the same time.*

e. *The object is sitting at the center of curvature of the mirror and at the incoming side focal point of the lens. So I guessed that this setup would send an image off into infinity. Sure enough. . . The mirror produces an image the same size as the object and located at the same position as the object but inverted. The image of the mirror is still at the focal point of the lens. Rays going into the lens from this point come parallel, so the image produced is at infinity (so. . . aliens can see it? . . . what does infinity actually mean?) The connection with lighthouses is that you put an intense light source in the object place, and you get light streaming off into infinity. . . or at least as far as a wary ship captain's eyes. PS I've had a fascination with photographing lighthouses since I was a kid (they make nice stark images). . . I've seen this diagram at a few historical sights. . . but I think the scale is larger. . . does my memory serve correct?*

Chapter 27: Optical Instruments

WARM-UPS

1. Nearsighted people can see objects clearly if they are close, but not if they are far away. Does a nearsighted eye have a focal length that is too long or too short?
Section 27-2

This is a relatively easy question that makes a good beginning to the discussion of vision. Most students can find the answer, but have not given as much thought to it as they might. I find that this makes a good launching point for a discussion accommodation, and of the physiology of the eye. For instance, I often move from here to a calculation of the range of focal lengths required for a person to see clearly from infinity to close range. From here, it is easy to make the transition to a discussion of corrective surgeries and corrective lenses (in combination with Question 27-2).

2. Reading glasses are supposed to give a virtual image of an object held at arms' length further out where the farsighted person can see clearly. Estimate the focal length of a reading glass lens such that a newspaper held at 25 cm from the eyes will image at 40 cm from the eye.
Section 27-2

Along with the last problem, this makes a nice introduction to corrective optics. Further discussion can include the difference between glasses and contact lenses. This also creates a good time to stress that corrective lenses produce virtual images, and to introduce the notion of lens power. I then turn this discussion around and ask the class about the range of focal lengths available in reading glasses. Some students may have noticed that in most drugstores, it is possible to buy nonprescription reading glasses with power up to 3.25 diopters. Why? What far point does this correspond to?

3. In a simple refracting telescope, there are two lenses, the objective lens and the eyepiece. Please explain what each of these lenses does.
Section 27-5

Students will have read the description of the telescope, but many will not have really thought about the purpose of the two lenses. This question forces them to dig a bit deeper. You can bring up the difference between the telescope and the microscope and discuss the two systems in parallel, using students' answers to this question as a guide. This allows a discussion of the two systems from a practical viewpoint, that is, where is the object being viewed? Where should the final image be focused? etc.

4. You wish to make a telescope using lenses that you have around. Let's say you have a magnifying glass and a pair of reading glasses. Estimate the highest magnification you can get.
Section 27-5

This is a great way to bring students into thinking about what is possible, and what the devices around them really are. For instance, even after discussing reading glasses, calculating the focal length required, etc., most students are still inclined to consider them as "static" objects. Sure they are converging lenses, but they are used as reading glasses. This question opens them up to seeing the glasses as lenses. It also is a great opening to discussing the practicalities of a telescope. Most students can come up with values, say, 25 cm for the glasses and 5 cm for the magnifier to get an angular magnification of 5 for the *telescope. But how does it get put together? I usually move to a discussion of which lens goes where and how far apart they must be. Students will come up with a host of interesting questions ranging from "Why can't you use a piece of flat glass with $f = \infty$ to get an infinite magnification" to "Don't you need a cardboard tube?"*

PUZZLES

"Phishics"

A person with good vision finds that she cannot focus on anything underwater. However, plastic goggles with "lenses" that are flat plastic disks allow her to see the fish clearly. Please explain how this can be. (Hint: This has nothing to do with salt or chlorine in the water.)
Sections 27-1 & 27-2

About half of the students get this one largely correct, as in responses d-f. That is, they understand that it is the difference in index of refraction between air and water that makes the difference. Note, though, that some of these students misinterpret the importance of this difference; they refer to the possibility that this could be "compensated" somehow, or you could get "used to" the difference (responses d, e). Other students have no idea, and create a variety of justifications (responses a-c). I have used this question without the "hint." In this case, a large number of students opt for this "simple" answer rather than thinking about optics at all. In class, a straightforward generalization of this question is to consider the effect of immersion in water (or other liquids) of other optical elements or systems.

Here are some sample (unedited) student responses that provide useful starting points for discussion:

a. *The plastic lens in the goggles is a single lens representing everything in the water. Objects are presented larger than the real object and often closer. Without the one single lens in the goggles, the water against human eye lens acts as several lenses perpendicular to every point on the curvature of the eye lens. This effect causes refractions at each point, therefore the image diverges all over and is very distorted.*

b. *The flat goggles keep the focal point at the same place as it would be in air, so the water does not distort the person's vision or change her focal point.*

c. *Because if you didn't have the goggles on the water would be right against your eye. This acts like contacts and distorts your vision*

d. *The fact that you have lenses in front of your face does nothing. It is the fact that there is a volume of airspace kept between your eyes and the water. This keeps the index of refraction the same as what the eye is used to and so images formed can be clear. Without the goggles the index of refraction has changed, to whatever it is when the eye is submerged in water, and your brain cannot compensate, so the objects are blurry. Do you think that fish are blurry out of water?*

e. *The eye is designed (or naturally trained) to accommodate for incoming light from an atmospheric medium instead of water. Since the water has a different index of refraction than air the incoming rays are not manipulated correctly upon entering the lens. The plastic flat eyepieces allow for the air to cornea environment. Although the incoming rays will be manipulated prior to entering the eye they will be parallel and will represent their 'natural' appearance due to eyepieces being flat.*

f. *This has to do with the fact that water has a different index of refraction than air. The incident angle coming in is refracted at a different angle so that the image is not formed exactly at the location of the retina, therefore the image is not sharp. When one puts on flat lenses in front of their eyes, there is once again the same index of refraction of air on the outer surface of the eye.*

Chapter 28: Physical Optics: Interference and Diffraction

WARM-UPS

1. The picture below shows light from a narrow laser beam as it appears on a screen after passing through TWO narrow slits. Each small division represents a distance of 1 cm.
Please examine the pattern and answer the following questions.

What is the distance between two successive minima due to interference between the slits?
What is the distance between two successive minima due to diffraction at the slits themselves?
What would change if the separation between the slits were reduced?
What would change if each slit were made wider?
<u>Sections 28-2 & 28-4</u>

The question is asking the student to pay careful attention to the narrative in the chapter on interference. Just looking at the formulas won't do. The student is also asked to interpret the intensity graphs and relate them to the image.

The distance between the first interference minimum on the right and the corresponding minimum on the left is a little more than 2 cm. The separation of the first few interference minima is about 1 cm.

The central diffraction maximum is about 24 cm wide, which is also the distance between the first diffraction minimum on the right and the corresponding minimum on the left.

If the separation between the slits were reduced, the distance between interference minima would increase.

If the slits are made wider, the width of the diffraction maximum would decrease, i.e., the diffraction minima would move closer together.

2. The picture on the left shows what happens to a narrow laser beam after it passes through a small, round pinhole. The same happens to light passing through your pupil. In your textbook, find the criterion for resolving two light sources. Use the criterion to estimate at what minimum distance your eye would be able to resolve the two headlights of an approaching car. (Estimate the diameter of your pupil to be 2 mm.)
<u>Section 28-5</u>

Here we are looking for the Rayleigh criterion for diffraction limit on a circular aperture. The minimum resolvable angular separation is given by $1.22\lambda/D$.

For visible light the wavelength is about 500 nm and the aperture D is the pupil diameter of 2 mm. The minimum angular separation of two sources is thus about 3×10^{-4} *radian.*
Estimating the headlamp separation to be about 2 meters we get an estimate of the minimum car distance to be about $2\text{m}/3 \times 10^{-4} = 6$ *km.*

3. Why does a soap bubble reflect virtually no light just before it bursts?
Section 28-3

The light reflected off of the thin film of water that makes up the soap bubble skin is a superposition of the light rays reflected on the inner and outer water/air interfaces respectively. The ray reflecting off the outer skin surface (air/water interface) undergoes a half-cycle phase shift. The ray reflecting off the inner skin surface (water/air) interface reflects with no phase shift. As the bubble skin gets very thin, the two rays reflect at virtually the same point in space. The only phase difference between them is due to the phase shift on reflection. The rays are out of phase by 180° and the bubble looks dark.

4. Rock specimen slices are often 30 micrometers thick. Approximately how many wavelengths of visible light does that represent? What additional information would you need to be able to answer this question exactly?
Section 28-1

The term "visible light" covers a range of wavelengths. Taking the average wavelength to be 500 nm, the answer is 60. To give an exact answer, the exact wavelength would have to be specified.

PUZZLES

"One Polarizer, Two Polarizer..."

Two crossed polarizing filters (oriented with their axes at right angles to each other) almost completely annihilate a light beam (initially unpolarized). If a third polarizer is placed between the two, the light intensity usually increases. Why is that? How would you orient the middle polarizer to maximize the intensity of transmitted light? What percentage of the original unpolarized beam would pass through such a setup?
Section 28-5

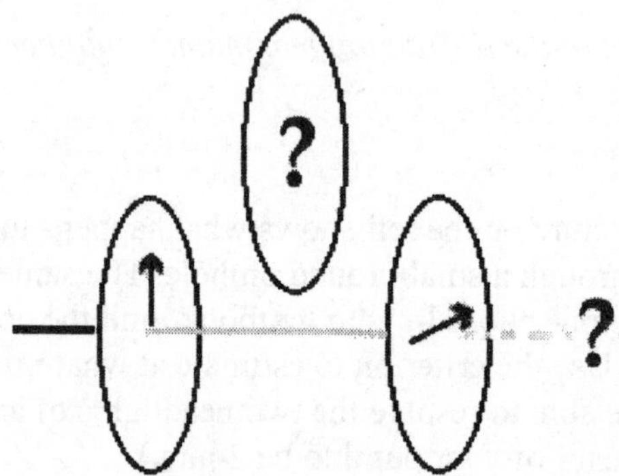

A polarizer will pass the components of an electromagnetic wave that line up with the polarizer's preferred direction. If the light is unpolarized, only a small fraction of light energy will be transmitted. The components of the wave that are orthogonal to the preferred direction are absorbed by the polarizer. The (linearly polarized) light that emerges from the polarizer gets totally absorbed by the second polarizer if its preferred direction is perpendicular to that of the first. When a third polarizer is inserted between two crossed polarizers it may well happen that the linearly polarized light, emerging from the first polarizer, has a component that can pass through. The passed component will not be perpendicular to the second polarizer and thus part of it will pass through.

Chapter 29: Relativity

WARM-UPS

1. Einstein's theory of special relativity predicts that the rate at which a moving clock runs will be slower than the rate of the same clock when stationary. Suppose you are shut in a rocket module which is moving through space at half the speed of light. Would you be able to observe the "slowing down" of the clock you brought with you from Earth?
<u>Sections 29-1 & 29-2</u>

The primary goal of this question is to start getting students into the swing of thinking relativistically. Classroom discussion can move from here to a few problems with obvious similarities (e.g., would you observe dilation of a ruler you brought with you) to the analogous questions concerning tools observed in other reference frames. From this point it is natural to move on to simple calculations.

2. Suppose you had a car engine that would convert the rest energy of ordinary water into kinetic energy of the car. Estimate the amount of water needed to power an average car for the lifetime of the car. (Use reasonable estimates for the number of useful years and for average yearly mileage.)
<u>Section 29-6</u>

Most students realize that the famous mass-energy equivalence produces astonishing amounts of energy from relatively little mass. This problem helps put specifics to that. From here it is straightforward to introduce a discussion of the small mass differences available in actual nuclear processes such as fission of uranium and fusion of deuterium. Ask students to redo the calculation estimating the amounts of these two materials that must be on hand to power the same car through the same distance. Incidentally, for many students the most difficult part of this problem will be the "review"; the estimation of how much work a ca does during its lifetime.

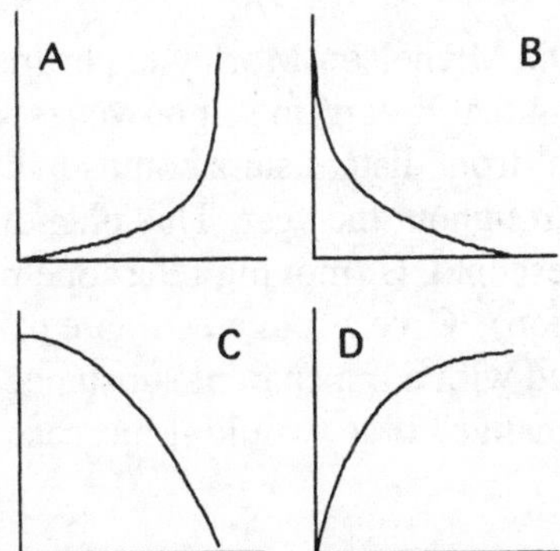

3. According to the theory of relativity the momentum formula p= mv has to be modified with the factor gamma. Which of the above sketches of p vs. v/c best agrees with the relativistic momentum formula? Why?
<u>Section 29-5</u>

This question helps connect the mathematics to the physics by asking students to think about the function at hand in simple, graphical terms. For the same reason, it is useful to ask the class what variables the other graphs might describe, e.g., graph D could represent velocity vs. momentum, or C could show the length contraction. In addition, this question introduces the topic of relativistic momentum, and provides a bridge to the distinction between mass and estimated mass.

4. Estimate how fast you would have to run to have a relativistic mass of 1000 kg.
Section 29-5

Wis acre students will put down "approximately c!" That is okay with me, but I will still hold their feet to the fire and demand a better answer. This question rounds out the discussion of relativistic mass, and helps give students a feel for how rapidly gamma diverges as v approaches c. In the classroom, this can be followed up with calculations of all the other relativistic variables. Students are particularly interested to be reminded, at this point, to an outside observer they would be not only much "heavier" but also much skinnier.

PUZZLES

"Ether,... Or"

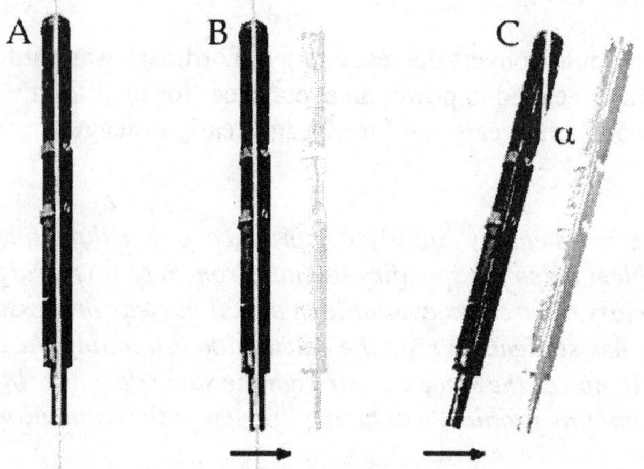

In an attempt to explain the null result of the Michelson-Morley experiment, a proposal was put forward that the Earth is dragging ether, the medium that is carrying light waves, along with it. This explanation did not agree with the observation that light from distant stars comes in at an angle which depends on the velocity of the Earth and which changes throughout the year. This phenomenon is called the aberration of starlight. See the figures: A (stationary telescope), B (moving telescope pointed at the star), and C (moving telescope adjusted for the aberration). Give a rough estimate of the magnitude of the aberration angle alpha. If the telescope tube were filled with a transparent substance of a high index of refraction, would you expect the aberration angle to change? If so, would it increase or decrease?
Section 29-1

This problem has two distinct purposes. The first is to get the students thinking "shifted" tothe realm in which the speed of light makes a difference. Even though it does not require a relativistic calculation, it does require students to work with the ration v/c. It is instructive to students to see that there are "classical" effects that depend on this parameter, and that they have visible consequences. The second value is historical. In addition to disproving the ether drag hypothesis (as discussed in the question, it is worth pointing out to the class that this effect was first observed by James Bradley in 1727. In fact, an understanding of aberration allowed Bradley to improve on the known value of c from his data.

Chapter 30: Quantum Physics

WARM-UPS

1. In your own words, explain what happens in the photoelectric effect.
Section 30-2

Many students read the section about the photoelectric effect with the same eye that they read other parts of the book. That is, "what part of this is needed to solve test problems?" As a result, they miss the historical significance of this effect in motivating the quantum theory. This question gets them started on an understanding of this aspect, and the classroom discussion can complete the process. Since quantum theory is so counterintuitive, its development makes a particularly good lesson in "how science works." Discuss the kind of evidence required before scientists will explore "outlandish" theories, and the level of verification required for such ideas to hold up.

2. A certain radio signal has a wavelength as long as your arm. Estimate the energy of one photon of this radiation.
Section 30-2

The calculation here is relatively simple, and most students will get it right. The importance here is to introduce the photon idea. The classroom discussion can then move on to a much more complex topic: the relationship between this picture and the wave picture discussed in Chapter 25. Remind students how they calculated intensity before, and how they can calculate photon flux now. Specifically link the two ideas in class. Work through the relationship between numbers of photons and electric field strength, etc. Finally, estimate the number of photons emitted per second by a 100 W lightbulb.

3. Estimate your de Broglie wavelength when running as fast as you can.
Section 30-5

Like the previous question, this introduces a simple quantum calculation, and focuses on the comparison to human scale. Most students will not have difficulty doing the problem, though they may be unsure about the answer since it is so far from the realm of human experience. It is useful in class to move from this scale to the atomic scale, asking for the wavelength for, say, an electron accelerated through 1,000 V. Moving back to a photon can reaffirm the notion that this wavelength calculation is exactly the same kind of operation as was done in the previous problem.

4. Estimate the uncertainty in the momentum of a hydrogen nucleus given that we know it is in a diatomic hydrogen molecule of length 0.075 nm.
Section 30-6

This question introduces another standard type of calculation. This time, though, the emphasis is reversed; we ask the student to do the atomic scale calculation, and the class discussion may include a calculation for an uncertainty at "human scale." (For instance, "We know you are in the room, what is the uncertainty in your momentum?") Many students will be surprised by the number that results from this calculation; they think they understand that small values come about for quantum calculation on large objects (such as the previous question). They expect a large uncertainty here. They may be "reassured" if you ask them to go on and calculate the uncertainty in velocity for the proton.

PUZZLES

"Uncertainty"

An electron passes through your apparatus at t = 0. You measure its position with accuracy Δx_0. How accurately can you calculate its position later on, at time t_1? (Hint: The uncertainty in the electron's velocity is related to the uncertainty in its momentum.)
Section 30-6

Faculty recognize this as the classic thought experiment about uncertainty, but students will not. It is worth having them work through this on their own to get an appreciation of it, and to get some experience working with quantum concepts. In class, the discussion can center on this issue, and can include a variety of extensions, including the one suggested by the figure, in which the uncertainty of the position is related to the wavelength of light used to probe the eledron. To extend this question, it is important to lead students to put in realistic numbers. Next, it is valuable to redo the exercise for a more classical case, e.g., a dust particle.

Chapter 31: Atomic Physics

WARM-UPS

1. What experiment suggested that atoms are not solid but that they may have a hard nucleus encircled by negative electrons?
Section 31-6

The experiment was performed in 1910 by Ernest Rutherford and his colleagues (one of whom was Hans Geiger, the inventor of the Geiger counter). They observed that electrons, impinging on a thin foil of gold, went right through the foil except for an occasional electron rebounding straight back. Rutherford and his team attributed the rebounds to the electron hitting a small piece of positive charge (the nucleus), situated in an otherwise empty space through which most of the electrons passed unobstructed.

2. A 5-mW laser pointer beam has a diameter of 3 mm. The wavelength of laser light is 680 nanometers. Estimate the number of photons per second that hit the laser spot on the wall.
Section 31-2

Each photon in the beam carries an energy of $hf \sim 3$ *J.*
The energy is reaching the wall at 5 mJ per second.
That represents about 2×10^{16} *photons per second.*

3. Niels Bohr suggested that an atom emits light as an electron, orbiting the atom, moves closer to the nucleus. How did Bohr explain the fact that spectra of chemical elements (such as hydrogen) consist of discrete colors rather than the whole continuous spectrum?
Section 31-6

This question is asking students to express the idea of quantization in their own words. While the answer is relatively simple, Bohr's idea was that only certain values of the electrons angular momentum are allowed, the student responses will reveal a variety of issues that have to be dealt with.

4. The human retina can respond to the light energy of a single photon! The wavelength of visible light is of the order of 500 nm. Compare the energy of a photon of visible light to that of a 100-gm rifle bullet traveling at the speed of sound (340).
Section 31-2

The ratio of the two energies is $E_{photon}/E_{bullet} = 6 \times 10^{-34}$

PUZZLES

"Isn't this Backwards?"

According to classical mechanics, as an electron shifts from an outer orbit to an inner orbit its speed has to increase. Thus the kinetic energy of the electron has to increase. Yet, according to Bohr, during such a process a photon will take away the excess energy. Where does the excess energy come from?
Section 31-6

The extra energy comes from the electrostatic potential energy change of the electron/nucleus system. As the electron shifts to a lower orbit it loses potential energy.

Chapter 32: Nuclear Physics and Nuclear Radiation

WARM-UPS

1. What aspect of short-lived radioactive substances make them a health hazard? What aspect of long lived radioactive substances make them a health hazard?
Section 32-3

Radioactive substances with short half-lives have a high activity. They are highly dangerous for a short time. The substances with long half-lives are not as highly active but they stay around for a long time and may get concentrated by various natural processes, e.g., by moving up the food chain.

2. The disintegration constant of technetium-99, an important medical tracer, is 10^{13} per second. Estimate the half-life of technetium-99.
Section 32-4

This is the first of two questions that ask the student to reflect on the meaning of the disintegration constant.

The disintegration constant λ *appears in the equation* $N = N_0 e^{-\lambda t}$ *which describes the number of radioactive atoms left after a time t if the starting number in the sample is* N_0.
The half life $T_{1/2}$ *is defined as the time interval during which half of the sample decays.*
Thus $T_{1/2} = -\ln(1/2)/\lambda = 7 \times 10^{12}$ *seconds ~ 200,000 years*

3. The activity of a radioactive sample depends on

i. the property of the radioactive MATERIAL and
ii. the size of the SAMPLE.

The material is characterized by the disintegration (or "decay") constant which states what fraction of a sample will undergo the process per unit time. The disintegration constant of radioactive carbon-14 is equal to 3.84×10^{-12} per second. Based on that information, how many counts per minute would you expect for a mole of carbon-14? Explain your answer. (Reminder: a mole consists of 6×10^{23} items.)
Sections 32-3 & 32-4

This is the second of two questions dealing with the meaning of the disintegration constant. The disintegration constant also appears in the expression $dN/dt = -\lambda N$, *where dN/dt denotes the activity of the sample in counts per second and N denotes the number of radioactive atoms present. dN/dt is a negative number since the number of atoms N is decreasing. Substituting the above carbon-14 data we get* 1.38×10^{14} *count per minute per mole.*

4. The half-life of carbon-14 is 5730 years. Living organisms, such as plants, continuously replenish carbon-14 in their system from the atmosphere. After they die, the carbon-14 content no longer is maintained. You are given a piece of wood whose carbon-14 activity is 25% of what you observe in living wood. How long has the wood been dead? (Explain in plain language how you obtained your answer by simple reasoning.)
Section 32-4

The activity of a radioactive sample is proportional to the number of radioactive atoms present in the sample. The activity of the sample in the question is ¼ of what it was when the sample was formed. The sample has undergone two halvings, so it must be 11,460 years old.

PUZZLES

"A Hot House"

Radium and Radon are both naturally occurring radioactive substances found pretty much everywhere. Why is Radon considered a much more worrisome environmental problem?
Section 32-7

While radium is solid, radon is a gas. Radium in building materials is much less likely to get into people's bodies than the airborne radon. Both radium and radon are alpha emitters.